If you thought Alexander Graham Bell invented the telephone think again. This book starts with the invention of the telegraph in 1840 and follows the developments right through to the introduction of mobile phones in 1980. It is a tale of skulduggery, court cases and many supposed experts saying "It will never catch on".

When he first saw the telephone American President Hayes summed up the sentiment in 1876:

> *"That's an amazing invention but who would ever want one?"*

Many fortunes were made but even more lost and as we shall see there is nothing new with Google, Spotify or Fake News, they all had their forebears over 100 years ago.

The last book on the History of the Telephone was written in 1910 by Herbert Casson, it seems it is about time a new one was written so read on….

John Lucas April 2021
john@itwillnevercatchon.com

Table of Contents

Prologue 3

Chapter 1 – The Telegraph 6

Chapter 2 – Wiring the World 15

Chapter 3 – The Fax Machine 27

Chapter 4 – The First "Inventor" of the Telephone – Antonio Meucci 30

Chapter 5 – The Second "Inventors" of the Telephone – Graham Bell and Elisha Gray 38

Chapter 6 – Roll-out of the Telephone 55

Chapter 7 – The Automatic Telephone Exchange 69

Chapter 8 – Phone Kiosks 83

Chapter 9 – Telephone Instruments 88

Chapter 10 – Telephone Numbers 93

Chapter 11 – The White Heat of Technology 97

Chapter 12 – Competition 108

Chapter 13 – The Mobile and Internet Revolution 120

About the Author – John Lucas 128

Prologue

The Battle of Waterloo was fought in Belgium on Sunday 18th June 1815 but it was not until the following Wednesday that news of the victory was received in London, a hundred years later news of the battles in the trenches of Belgium were relayed by telephone in a few seconds and the newspaper reporters could report from the front line directly to their editors in Fleet Street. It is hard to imagine not being able to communicate with anywhere in the world but until 1865 it took over three weeks for a message to reach America from Europe and another three for the reply.

Today anyone in the world can call anyone else within a few seconds and at very little cost yet we forget that in living memory things were not as easy. In the 50's and 60's making a call was much more expensive and the networks less reliable. When I worked in Greece in the 1990's the telephone service was awful. It was suggested at the time that Athens did not win the bid for the 1996 Olympic games because the committee, when they visited Greece, found they could not call home so the games went to Atlanta instead. After the EU poured a great deal of money into the Greek telecoms infrastructure it improved out of all recognition and Athens were awarded the 2004 games. Without a reliable telephone service a country cannot function and today we would also add the need for internet as well.

So where did it all start?

Before 1800 the knowledge of electricity was limited to flying kites in thunderstorms or touching electric fish, clearly electrical storms had great power but no one thought the same phenomena could be used in more controlled and modest ways. Between 1800 and 1850 this all changed when the leading scientists of the day started to understand what electricity was, how it could be produced, stored and used. No one person can be credited with "discovering" electricity or it's uses just as you can't say who invented steam power, everyone has always known steam was a potential power source but it took a dozen or so bright engineers to develop a practical engine, each adding a few improvements of their own. In the same way many different scientists such as Voltare, Ampere, Maxwell and Faraday contributed to the understanding of electricity so by 1830 it was possible to use it in a practical way.

By 1850 enough was known about this fantastic new science that it started to move from the laboratory into devices which were of benefit to the public. The most dramatic of these was the telegraph and then the telephone both of which revolutionised the way the world communicates, does business or receives news.

The telegraph and telephone are derived from the Greek words; "tele" which means "far", "graph" is "written", whilst "phone" is "voice sound".

This book describes the history from the first telegraph messages right up to the present day when anyone can call anyone else in the world in an instant. In 2007 there were 1,261 million fixed line telephones in service but today (2021) only 915 million so perhaps now is the time to write it's epitaph. This demise is of course due to the rapid rise of the mobile phone which has gone from zero in

1980 to over 8 billion today, there are more mobile phones in use than people on the planet (8 billion against 7.8 billion). This is a statistic I have trouble getting my head round but as we will see in this story I am in good company, a great many people thought the telegraph and then the telephone were not worth investing in.

When the telegraph was first proposed the US President and his cabinet watched a demonstration but went away unconvinced and the New York Times called the telegraph;
"superficial, sudden, unsifted, too fast for the truth".

When the telephone was first offered to Western Union, the major telegraph company of the day, it's Chairman said:
"What use could this company make of an electrical toy?"

In 1879 Sir William Preece of the British Post Office, when asked whether the telephone would be popular in the UK replied:
"I think not, I fancy the descriptions we get of its use in America are a little exaggerated; but there are conditions in America which necessitate the use of instruments of this kind more than here. Here we have a superabundance of messengers, errand boys, and things of that kind."

The first mobile phone:
*"Too expensive – **It will never catch on**"*

I seem to remember I said that in 1986 but in my defence I was not the only one.

Chapter 1 – The Telegraph

This may seem a strange thing to start with when writing a book about the telephone but I would like to tell the story of a world famous American portrait painter who, being one of the foremost artists of the day, was even commissioned to paint the Presidents portrait in 1820.

A critical moment in his life was an incident in 1825. He was in Washington painting the portrait of the Marquis de Lafayette who was a leading French supporter of the American revolution when a messenger arrived from his father which read "Your dear wife is convalescent". The next day he received a second message to say his wife had died. By the time he got home, which was only 300 miles away, his wife had been buried. He was shocked that he knew nothing of her failing health until it was too late and he resolved to explore a means of rapid long distance communication.

He visited Europe in 1830-32 to work on a painting in France called "Gallery of the Louvre" and on the return crossing all the ships first class passengers discussed the issues of the day around the dining table, as the journey lasted many weeks they had plenty of time for discussion.

One evening a fellow passenger by the name of Dr Charles D. Jackson had begun a conversation about the curious properties of electricity and recounted his experience of a memorable experiment at the Sorbonne where he had watched as an electric spark fizzed four hundred times about the lecture hall in an instant.

Another passenger then asked whether a wire would obstruct the passage of an electric current. Jackson replied, 'No, Benjamin Franklin has demonstrated that electricity travels at once through any length of wire.' The painter then said 'If the presence of electricity can be made visible in any part of the circuit, I see no reason why intelligence may not be transmitted instantaneously by electricity.' He later claimed that as he spoke this sentence an idea appeared in his imagination. He excused himself from the table and went out on deck to scribble his thoughts in a notebook and he started to formulate his ideas.

His name? Samuel Morse.

Jackson's conversation had set him thinking there might be a solution to the problem of communication which had been bugging him since his wife's death.

For nights afterwards he was unable to sleep for excitement, being locked away on a ship was the ideal environment to formulate his ideas but one gets the feeling his enthusiasm and excitement irritated his fellow passengers. However, several of them came to his aid in court when he needed to prove the telegraph was his idea alone as those impromptu talks in the dining room of the ship were to be the subject of litigation for many years to come.

Dr Jackson, when he realised that there might be money to be made from the invention tried to claim a share of the spoils of inventing the telegraph. It seems Jackson tried a similar trick by claiming the prior invention of gun cotton and the use of ether as an anaesthetic, he lost both cases.

Morse knew little about electricity, after all it was a very new science and he was an artist not an engineer, his first designs were not very successful with the first prototype being far too complicated. Initially the electrical impulses were displayed on a crude printer using paper which darkened when a current was passed through it but the main issue facing Morse, and all other inventors, were the losses due to resistance when the length of wire was increased.

The breakthrough came when Morse added a relay as a repeater at intervals along the line which could extend the range infinitely.

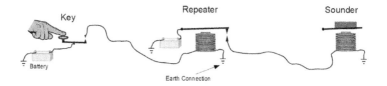

The diagram above shows the main components of Morse's telegraph. The sender taps the key which connects a battery to the line which in turn activates the electromagnet at the repeater. The electromagnet operates a switch which connects a battery to the next wire in the chain. So each length of wire has a new power connection and as many can be added as needed to extend the range. At the distant end an electromagnet activates a sounder, a crude buzzer, so the receiving operator can hear a noise when the key has been pressed. The use of a printer had been dropped. One important point to note is that only one wire is required between locations.

Progress was slow and Morse's finances were in a poor state as his artistic career seemed to be in terminal decline, perhaps because of his obsession with his new idea. Then in March 1837 a circular was published by the Secretary to the US Treasury who had been asked by the Government to investigate the possibility of what we now call the telegraph. This perhaps spurred Morse on, by now he had the support of Professor Leonard Gale of the University of the City of New York. With his help they managed to demonstrate a system which could send a message over a third of a mile.

One of the people to witness the demonstration was Alfred Vail, aged 30, who became a partner in the enterprise who was key to the development of the telegraph.

Samuel Morse

Alfred Vail

Vail was the son of Judge Stephen Vail who owned an iron and brass works in New York, Judge Stephen not only financed the development work but also made his factory, with its workshops, available to the venture. In return Alfred Vail was given a 25% stake in the business.

It seems that Morse was the ideas man and good at promotion but, like a lot of artists, was not practical. Vail on the other hand was much more reserved and practical so was the ideal partner in this enterprise.

By now Dr Jackson was making claims that he was the joint inventor of the telegraph, although he had a reputation as a serial litigator Morse still had to take time out to defend his case against Jackson whilst at the same time he and Vail set to work on the engineering.

Morse initially used the system to send numbers which could then be converted to text. Somehow during the development work this idea was replaced by a code using dots and dashes, although we call this Morse code it is not known if it was Morse, Vail or a collaborative invention. Vail was a very modest man and in a book he wrote later about the development of the telegraph he did not say who came up with the idea so perhaps it was his idea and we should call it "Vails Code".

This work was effectively funded by Judge Stephen Vail who was probably a little concerned where his money was going so on 11th January 1838 he was invited to the workshop and asked to write a sentence on a piece of paper which his son then tapped out in morse code.
A few moments later Morse entered the room with the words "A patient waiter is no loser".

This method of demonstrating the system, get someone, ideally a prominent individual, to write a sentence of their choosing on a piece of paper which is then transmitted to the other end where the receiver can "announce" the words became the format of future demonstrations. Just like a magic trick at the theatre except there is no magic. The Judge was now a convert and was keen to petition government for support.

The following month the system was demonstrated in Philadelphia and Washington where the good and great including the President came to see it in use. Unfortunately most visitors, whilst impressed saw it as a scientific curiosity rather than of anything which might be of use! Consequently no one came to invest in the enterprise, **it will never catch on** seemed to be the sentiment.

With this setback Morse went to Europe to try and gain financial backing and obtain patents, however he discovered there were rivals!

In 1833 Carl Fredrich Gauss and Wilheim Weber created the first telegraph which connected their observatory with the Institute of Physics in Göttingen, a distance of 1.2km. This demonstrated that it was possible to create a circuit with with a switch, a battery and a simple electromagnetic buzzer or indicator with which to transmit messages over a distance instantaneously. They did not have the backing to develop their invention commercially but a system was built in Munich and run along the first German railway in 1835.

Wheatstone & Cooke's
6 wire telegraph

Science Museum London

At the same time that Morse was developing his system in America, in England William Cooke and Charles Wheatstone had patented a six wire system with a small keyboard and a number of dials which indicated the letter or number being pressed.

This system had the obvious advantage of removing the need to code and decode the message however the poor quality of insulation at the time meant that having 6 wires, rather than one in the Morse system, was a major limitation making the system far less reliable.

Their system was used extensively on the railways in Britain but, because of the shortcomings of the 6 wire system, most only had two wires and were used to manage the running of trains.

Morse and Vail's efforts seem to have been in vain until in 1843 when they were at the point of giving up, a government grant of $30,000 (a million dollars in today's money) was finally awarded to them to build a line between Washington and Baltimore.
Building it took a great deal of effort with many design changes on the hoof. For example, initially the wires were

run underground but the insulation of the period was not up to the task so the telegraph pole was invented. Eventually on 24th May 1844 a line between the Supreme Court Chamber in Washington and Mount Clare railway station in Baltimore, a distance of 44 miles, was opened in front of a crowd with the message "What hath God wrought", a passage from the Bible selected by Miss Annie Ellsworth, the daughter of one of Morse's supporters.

Whether by chance or design the opening of the telegraph line coincided with the Democratic National Convention which was being held in Baltimore. There were two favourite candidates to win the nomination but neither could win a majority. Much to everyone's surprise a third candidate, James Polk who was only expected to be nominated as Vice President, was chosen as a compromise and Morse was able to flash the news to Washington gaining considerable positive publicity with the first ever example of breaking news.

Eleven minutes after Congress heard of Polk's success they were able to reply telegraphically: 'Three cheers for James J. Polk.' It would have taken a day for the news to have reached Washington without this new invention. So now the concept had been proven all that was needed was the capital to launch a service but where would the investment come from?

Chapter 2 – Wiring the World

By 1844 the telegraph had overcome its doubters and a "telegraph boom" followed with companies competing to spread their networks far and wide. These networks consisted of steel wires (not copper) and a message was sent over a single wire with an earth connection at each end completing the circuit.

Telegraph networks were expensive to build, but financing was readily available, especially from London and New York bankers. Having "instant communications" was seen as giving a significant commercial advantage, just as today no company could manage without a good internet connection then access to the telegraph was just as important.

Morse's telegraph became the European standard in 1851 however Britain kept on with the Cooke and Wheatstone system for some time after that. By 1852 the US had 20 telegraph companies with 23,000 miles of wire between them and in the UK there were 2,200 miles of wire.

Because of the special skills required to send or receive Morse code messages they were sent between two telegraph offices. The customer wrote their message out on a form with the address of the recipient and took it along to the telegraph office, a fee was paid and the message was sent to the nearest office to the recipients home or office and a messenger took the message on the last leg of its journey. Large companies and governments had their own networks so could communicate quickly between their offices.

It cannot be overestimated the effect this "instant" messaging had on society. The ability to communicate vast distances changed peoples perception of the world.

Just as today there are "disruptive technologies" driven by the internet such as Uber or Kindle so the telegraph had an impact. It took 10 days for the Pony Express to deliver a message from the East to the West coast of America compared with a few minutes by telegraph which completely killed the Pony Express business.

One significant positive effect was the introduction of weather forecasting, if a storm hit one area the message could be relayed to others in its path thereby warning them of the danger, this saved the lives of many seamen who, warned of gales heading their way, did not put to sea. Before then the arrival of a storm was always something of a surprise.

After a murder was committed in Slough the assailant ran to Salt Hill Station and escaped on a train to Paddington. A telegraph message was sent ahead giving the details "He is in the garb of a Quaker, with a brown great coat on, which reaches nearly down to his feet; he is in the last compartment of the second first-class carriage". He was arrested at Paddington station and subsequently was tried and executed. The first known example of the telegraph being used in a murder case. The papers commented "No man could outpace electricity, however fast he ran."

The American Civil War of 1860-65 demonstrated the power of high speed communications. The Union Forces network carried 6.5 million messages during the war and they built 15,000 miles of line. General Grant wrote that he had "held frequent conversations over the wires" about strategy with Edwin Stanton, some lasting two hours, Stanton was Lincoln's right hand man who organised the logistics. General Sherman also recalled the "perfect concert of action" between his forces in Georgia and General Grant's in Virginia. "Hardly a day intervened when General Grant did not know the exact state of facts with me, more than fifteen hundred miles off, as the wires ran."

In contrast the Confederates in the south used the telegraph in only the most limited fashion.

Running wires between towns and cities was quite straight forward, crossing the sea posed a few more problems. In August 1850, having earlier obtained a concession from the French government, the English Channel Submarine Telegraph Company laid the first line across the English Channel.

It was simply a copper wire coated with gutta-percha (a natural rubber) without any other protection and was not successful. However, the experiment served to secure renewal of the concession, and in September 1851, a protected core, or true cable, 24 miles in length with an outside layer of steel for strength and the conductor inside insulated again by gutta-percha, was laid by from a government hulk, *Blazer*, which was towed across the Channel.

It worked and Britain was finally connected with Europe.

In 1853, more cables were laid, linking Great Britain with Ireland, Belgium, and the Netherlands. The Electric & International Telegraph Company completed two cables across the North Sea, from Orford Ness, Suffolk to Scheveningen in the Netherlands which were laid by *Monarch*, a paddle steamer which became the first vessel with permanent cable-laying equipment. Many of the early cable manufacturers were rope or steel cable makers who adapted their techniques for this new market.

Whilst being able to send and receive messages almost instantaneously was a benefit, especially in geographically large countries such as America, or between the UK and Europe the ability to communicate instantly between America and Europe or Australia would be revolutionary. A message sent from Europe to America took, even on the fastest steam ship, 15 days to get there, the answer another 15 days, if only a telegraph wire could be laid across the Atlantic then the 30 days to get a reply could be reduced to seconds.

The financial rewards would be huge but so were the risks, laying a cable 2,600 miles under the sea was a considerable technical and engineering challenge (wireless had not been invented then so it had to be a cable). The first person to attempt this was an American financier Cyrus West Field who had made his fortune in the paper industry, he set up the Atlantic Telegraph Company in 1854.

Apart from the issue of making a cable physically strong enough to be able to withstand the stresses of tides, sea currents and the weight of the cable itself the big problem was how to overcome the resistance of a 2,600 mile long cable so a pulse applied at one end could be detected at the other. It would not be possible to fit Morse's repeaters under the ocean.

This issue became a sorry tale of competition between the so called experts working on the project, the first of which, William Thomson, born in 1824, had done a great deal of research into the theory of long distance telegraph lines and was probably the worlds leading authority in the field but he was regarded as the junior to Wildman Whitehouse, born in 1816, who had trained as a medical doctor but had become an electrical experimenter and enthusiast. One gets the feeling he was also good at blowing his own trumpet. Whitehouse was appointed Chief Electrician on the project.

In 1857 the first attempt was made to run a cable from Ireland, however it broke repeatedly. The problems were largely due to difficulty controlling the tension as the cable was played out. A new mechanism was designed and successfully trialled in the Bay of Biscay in May1858 so another attempt at crossing the Atlantic was made.

Cyrus West Field

William Thomson aged 22

Wildman Whitehouse aged 40

The weight of 2,600 miles of copper cable which had to be stored on deck, was too much for one ship, the considerable weight so high up would make it unstable. The solution was to use two ships each of which were loaded with 1,300 miles of cable, they met mid Atlantic, spliced two ends together and then set off in opposite directions, one to Trinity Bay in Newfoundland and the other to Valencia Island in South West Ireland. Even with only half the load the ships were quite unstable and nearly floundered before they started to lay their cables.

Thomson's solution to the resistance problem was to have a very sensitive detector at the distant end and he designed an ingenious device (a mirror galvanometer) to perform this task. He was on one of the cable ships and throughout the voyage checked the operation by receiving messages sent from the other cable head to the end of the cable in the ship using his sensitive detector and thereby making sure the cable had not broken.

Whitehouse favoured using a very high voltage which showed he had a lack of understanding of the principles of electricity, particularly of insulation.

When the cable laying was completed Whitehouse connected his own apparatus and proceeded to make test transmissions. When these were not very successful he increased the voltage such that after 732 messages were transferred it failed altogether, but not before Queen Victoria sent one to the US President on August 16th 1858 which expressed a hope that the cable would prove;

"an additional link between the nations whose friendship is founded on their common interest and reciprocal esteem."

The President responded that,

"it is a triumph more glorious, because far more useful to mankind, than was ever won by conqueror on the field of battle. May the Atlantic telegraph, under the blessing of Heaven, prove to be a bond of perpetual peace and friendship between the kindred nations, and an instrument destined by Divine Providence to diffuse religion, civilization, liberty, and law throughout the world."

The company, on discovering the cause of the failure, dismissed Whitehouse and William Thomson took over the role of Chief Electrician. He went on to a very distinguished career developing the first and second laws of thermodynamics and became Lord Kelvin in 1866, the absolute temperature scale is named after him.

The exchange of messages between Queen and President very publicly proved the concept of a transatlantic cable although some doubters said it was all a hoax which was covered up by now saying that cable did not work any more. "Fake News" is nothing new. But there were enough investors willing to fund another attempt which was made by the company in 1865 by which time undersea cables had been laid in the Mediterranean and elsewhere. Whilst not in such a hostile environment as the Atlantic, the experience had improved cable design and manufacture.

The Great Eastern steamship which despite being the largest ship in the world with space for 4,000 passengers had not been a commercial success so after only seven years was adapted to be a cable laying ship. It was able to carry the whole 2,600 miles of cable thus avoiding the need to make a join mid Atlantic.

The cable was made using high quality copper conductor insulated with gutta-percha and surrounded by steel wires to provide tensile strength. It set sail from Valentia Island in the south west of Ireland on 15th July 1865. Unfortunately after just over 1,000 miles the cable snapped and was lost! Unperturbed the company raised more money and tried again almost exactly a year later and this time the cable was laid successfully and the first messages sent.

What is perhaps more remarkable the Great Eastern then sailed back and managed to recover the end of the broken 1865 cable using a grappling hook towed behind the ship in the position recorded when it was lost. The crew then managed to splice a new length on the end and deliver that to Newfoundland as well.

So by the end of 1865 there were two cables across the Atlantic and they both gave a number of years reliable service.

More cables were laid in 1873, 1874, 1880, and 1894 increasing the transatlantic capacity now the business case and technology had been proven. Although Thomson had a good understanding of the principles of long line transmission it was not until discoveries decades later improved the speed at which messages could be sent but each of these first cables could carry approximately 8 words a minute.

All the cables in these attempts as well as the ships that laid them were British and this expertise made sure that British submarine cable systems dominated the world. This was set out as a formal strategic goal, which became known as the 'All Red Line'. In 1896, there were thirty cable laying ships in the world and twenty-four of them were owned by British companies. British companies owned and operated two-thirds of the world's cables. By 1923, their share was still over 40%.

During World War I, Britain's telegraph communications were almost completely uninterrupted, while it was able to quickly cut Germany's cables worldwide. This cutting of an adversary's communication links has recently been highlighted again with the assumption that, should world war break out, the fibre optic cables that we rely on for communications throughout the world would be severed, leading to the collapse of the Internet. The British Government have recently decided to build special ships for the Navy to protect these assets but how you stop a foe cutting a cable anywhere along a 3,000 mile length is anyone's guess.

The UK also suffers from accidental cable cuts as the coastal waters around the UK are quite shallow (what is known as the continental shelf). If a large ship is expecting a storm the crew drop anchor and wait for the bad weather to pass. The wind pushes the ship which drags it's anchor along the sea bed, should it pass a cable it is likely to break it. This happens perhaps once a year to a cable around the British Isles and when it does the cable ship needs to find the two ends and splice them together. Not a five minute task.

Across the Atlantic where the water is much deeper the cables do not sit on the sea bed but float at their natural depth based on their buoyancy but well out of the range of anchors.

Today the most valuable telephone asset is the network of cables and ducts, cable TV companies were the first real competitors to BT's network outside the city centres and could only justify the expense with the TV packages giving much higher returns. The telegraph was no different, almost all the costs were in running out the wires, poles and repeaters across the country or undersea which was phenomenal. What if two or more messages could be sent down the same wire the value of the assets would double. This became the holy grail of scientists throughout the world during the next two decades.

So by the 1850's "instantaneous" electronic communication was in place with the infrastructure to deliver it being rolled out across the world. Just to put this in some sort of historical perspective the American west was still very much the "Wild West", it would be another 20 years before Custers last stand (1876), Britain started to transport criminals to Australia (1850) and the British fought wars in the Crimea (1854/56) and in India (Indian Mutiny 1857).

One early adopter of the technology was Paul Reuter who initiated his news service between Aachen, Germany and Brussels in 1851. Next he set up a news wire agency in London which delivered commercial news to banks and business users. The agency then expanded by selling subscriptions to newspapers who were particularly interested in receiving international news before their rivals. Reuters are now the leading news agency in the world.

The telegraph continued to be used until recently as Telex and Telegram services. Telex was used as a reliable, written means of communication between companies and banks for financial transactions, it only went out of use with the arrival of the modern fax machines in the 1970's. Telegrams were a continuing use of the telegraph network, the messages were now sent using a telex machine but the sender still had to write their message out on a form and give it to the Post Office clerk and at the far end it was printed out and delivered by a rider on a small motorbike.

Between the end of the Second World War and the 1960's if you wanted to become a GPO engineering apprentice you first had to spend a year as a telegram despatch rider, perhaps this was seen as a form of selection.

Very few people had their own phone before 1950 so a telegram was the most common way to deliver important messages quickly. My father was called to an interview for a job in 1950 by telegram and sadly many families found out the news of their loved ones deaths during both World Wars in the same way.

But many thought the next step was to be able to talk rather than send coded messages but strange as it may seem the fax came next!

Chapter 3 – The Fax Machine

We now think of the fax machine as a 1970's invention which for about 20 years was an essential tool of any office until it was made almost redundant by the internet. It was in fact invented by Alexander Bain in 1842 although the first practical machine, called the Pantelegraph, was designed by Giovani Casell in 1856, 20 years before Bells telephone. It's name is a combination of Pantograph, a machine that copies words or drawings and telegraph, the network used to carry them.

Caselli was initially ordained as a priest, however by 1849 he had moved to Florence and was teaching physics in the university where he also carried out research into various aspects of electricity and in particular the telegraph.

His transmitter consisted of a pendulum which dragged a stylus across the document to be sent which was written in non conducting ink (or paint) on metal foil. The insulating ink disconnected the stylus from the foil as it passed over. After each swing the foil "paper" was moved on a fraction.

The receiver had a similar pendulum and the transmit and receive pendulums were started together. Whenever the transmitter stylus detected a mark the stylus on the receiver blackened some paper impregnated with a chemical (potassium ferricyanide which reacts to a current passing though it). Slowly as the pendulums swung back and forth and the paper advanced after each swing a facsimile of the first drawing appeared at the far end.

Several scientists had been trying to perfect the Pantelegraph, the biggest issue to overcome was synchronizing the pendulums. Caselli solved this with a clock mechanism at either end which were kept in step thereby making his system far superior to any competitors.

Obviously having to draw or write the document on foil was quite a limitation so the machine was mainly used to transmit signatures to verify banking transactions, or as a "novelty" at exhibitions.

The prototype system was demonstrated to Leopoldo II, Grand Duke of Tuscany in 1856 who sponsored further development.

In 1860 Napoleon III saw a demonstration of Caselli's pantelegraph and placed an order for the service within the French national telegraph network that started the next year. With French support and funding a test was done successfully between Paris and Amiens with the signature of the composer Gioacchino Rossini as the image sent and received, a distance of 140 km. A further successful test was carried out between Paris and Marseilles, a distance of 800 km. French law was enacted then in 1864 for a fax document to be officially accepted in a transaction.

I am sure many people looked at the Pantelegraph and said, "**it will never catch on**" and they were probably right. There were many different versions created over the years but it did not become a success until the Xerox Corporation patented a new design in 1964 on which all modern machines are based.

But after this unexpected interlude let us return to the main story, the telegraph was taking over the world and the big research money was being spent on trying to send more than one message down the same wire, if you could do that the value of the network would double overnight. Few people were even considering the daft idea of being able to talk over these networks.

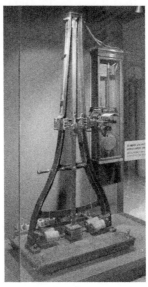

Museo Nazionale Scienza e Tecnologia Leonardo da Vinci

Chapter 4 – The First "Inventor" of the Telephone – Antonio Meucci

As every school child will tell you the telephone was invented by Alexander Graham Bell, as is often the case, what every school child tells you is wrong!

Firstly it is not strictly true that the telephone was "invented". Enough was understood about sound and electricity that it was just a matter of design, putting all the understood components together, to be able to carry speech across an electrical circuit.

The child's toy of two tin cans and a piece of string demonstrated the mechanical principles (carrying a vibration). The telegraph network had shown that a wire could carry a signal many hundreds of miles, now the challenge was to combine the two. There were three problems to overcome:

1. Convert air vibrations into a "vibrating" or oscillating current (the microphone)
2. Carry those current oscillations along wires.

3. Convert the oscillating current into sound (the earpiece or speaker)

Many scientists understood the issues and were working with the same basic understanding of electronics as described by Faraday, Maxwell and all the other great scientists.

The first "inventor of the telephone" was an Italian, Antonio Meucci (pronounced May-oo-chee), who was born in 1808. At the age of 13 he won a place at the Academy of Fine Arts in Florence whose teaching followed the example of Leonardo de Vinci, art and science together giving students a holistic view of the world. He was the youngest student when he started, so clearly a bright boy, with interests in chemistry, optics, electricity and mechanics. At the time he was studying Volta was the foremost scientist in Italy having invented the first practical battery in 1800.

Antonio's family were not wealthy and when he was 16 he had to start work to help the family's finances although he continued his studies on a part time basis. His first position was as a security guard for the municipal authorities which he performed well until in 1825 when, with three colleagues, he was asked to light the fireworks for a three day festival. On the first two days everything went to plan, the fireworks were set off at the right time and the crowds enjoyed the spectacle. On the third night one of the rockets went higher and further and then landed in the crowd injuring a spectator. The three firework lighters were arrested and although nothing was proven there were rumours that Antonio had "modified" the firework. From then on he was ostracised and accused of

other misdemeanours eventually spending three days in prison!
Despite this setback he finished his studies and was then given a job as a technician at the Teato della Pergola, one of the major opera houses in Florence where he lived in a room at the top of the building. He designed new lighting (candle based), improved the acoustics and was very highly regarded by the management of the Opera House. He also married one of the company's dress makers.

The company decided to open a new Opera House in Havana, and asked Meucci to become the project manager so he and his wife moved to Cuba in 1835. On the voyage which lasted 72 days he became interested in the work of the ships captain and in particular how to anticipate storms by the change in humidity. He made a type of hygrometer from a piece of whalebone, he just could not help inventing or improving things!

On arrival in Havana he oversaw the construction of the opera house though he was worried that the large auditorium would result in poor acoustics. To overcome this he diverted a river which ran close by under the floor to create a resonance chamber and reduce the volume of air in the building.

With his work on improving the sound, lighting (candles and mirrors mostly but he did introduce an electrostatic generator for flashes) he gained a reputation for his technical ability. The engineer in charge of the water supply to the city asked for help in improving the water quality. It was so bad people hired slaves to bring water from the springs in the hills for drinking and washing. Antonio analysed the water, discovered the cause of the problem and introduced filters containing lime and sodium which solved the issue. Another request was from the General in charge of the army. His soldiers had to send their swords, buckles etc to Europe to be electroplated as the humid atmosphere of Cuba quickly made them rusty, this was expensive, could Meucci help? He started an electroplating business which was quite profitable and employed 12 people.

He then experimented in electro shock treatment as a health cure. In one experiment he set up a circuit from the batteries in his basement to two small coils of wire inside pieces of cork, one for the patient and the other he held in the basement. I have no idea what his intention was but when the patient touched the wire at his end he screamed (it was 114v so it would hurt) and Antonio heard the noise in his piece of cork downstairs which got him thinking. He improved the design adding a funnel to concentrate the sound and stopped electrocuting his victims and just asked them to talk, it worked and it was the first one way telephone call ever made. Alexander Graham Bell was then only 2 years old.

Whilst his work was successful the Opera House itself was not making money, probably there were not enough opera lovers in Cuba to make it viable so in 1850 the theatre was closed and most of the company moved to New York.

Rather than return to Italy which was in turmoil at the time Antonio and his wife decided to move to New York as well despite neither of them speaking English. Whilst Antonio was hopeless with money his wife Esterre was not, she had carefully managed their finances during the 15 years they spent in Havana so on moving to New York they could afford to buy a house on Staten Island where Antonio set up his laboratory. Consequently he did not take up a position with the opera company.

He spent a few years improving the design of his "telegrafo parlante" as he called it (speaking telegraph). They also had a guest, Giuseppe Garibaldi the Italian general who was the founding father of modern Italy stayed with the Meucci's for six months in 1850/51 whilst he was in exile. He was co-opted into helping in a candle making business which Antonio had set up using a new process which made for brighter and cleaner light.

Around this time the first setback occurred, Esterre became bed ridden with rheumatoid arthritis so to ensure she could call for help Antonio fitted a telephone between her bedroom and his workshop next to the house. Several visitors to the house have since verified it's existence and that it worked. The worlds first two way telephone.

His next project was a new method of creating paper from wood for which he got backing from a newspaper company, he started to set up a paper mill in 1865 but the collapse of the economy at the end of the American civil war resulted in the financial backers going bankrupt. It is certainly true to say Meucci was a true polymath but he did not seem very commercially minded which, together with his poor English, hindered his ability to make much money out of any of these inventions.

Then on 30th July 1871 another disaster struck. He was on the Staten Island ferry when the boiler exploded killing 125 passengers, he survived but was badly injured and was bed ridden for a time during which his wife, or perhaps her carer, sold the telephone to raise a little money.

It is not recorded what Antonio said on discovering his telegrafo parlante had gone, perhaps his wife had become tired of his continuous talking about it. He managed to build a replacement and wrote up the technical description with a view to getting it patented. He visited a patent lawyer by the name of Thomas Stetson but as Antonio's English was so poor he needed a translator which probably did not help.

Stetson was not interested, he sent a letter saying, "Your telegraphing will have to be experimented with considerably before it will be ready for a patent. I advise making a good many experiments, to prove the reality of the thing. When you have got things to suit you, I wish to see the experiments, I will come down and stop a night at your house, if you will keep me and show me all about it." A full patent would have cost $250, money which Muecci did not have however he did manage to take out a "patent caveat" which gives temporary cover to an idea.

His next attempt at getting it backed was with a Vice President of the American District Telegraph Company, a Mr Grant, who accepted the designs and drawings and said he would make some telegraph lines available for a test. Antonio contacted him regularly during the next two years and was always fobbed off and eventually Grant said he had lost the plans and was not interested. Did he think "**It will never catch on?**"

It could also be the case that the designs were passed to their labs to test, the company were part of Western Union who eventually went into the telephone business as we shall see later.

When eventually Bell patented the "first" telephone Meucci tried to appeal. Bell had the best lawyers and there also seemed to be an element of skulduggery around the case. Whilst the entry in the patent records for Antonio's patent caveat of 1871 was there for all to see (he renewed it in 1872 and 1873) the patent office had lost all his drawings and plans, as a result he lost the case!

Antonio Meucci died in 1888, a combination of his poor English and poor promotion of his inventions mean that if he had solved the problems, which he probably had, then no one took much notice.

He was finally recognised in 2002 with a US House of Representatives bill which stated that "that the life and achievements of Antonio Meucci should be recognized, and his work in the invention of the telephone should be acknowledged" about 130 years to late but at least he is not forgotten.

His house on Staten Island is now a museum dedicated to Antonio Meucci and Giuseppe Garibaldi, see www.garibaldimeuccimuseum.com.

Antonio Meucci

Chapter 5 – The Second "Inventors" of the Telephone – Graham Bell and Elisha Gray

Two rival scientists were working on the problems at the same time; Alexander Graham Bell who everyone has heard of and Elisha Gray who is almost unknown. Interestingly they knew each other and may have had access to Antonio Meucci's notes which makes the following story all the more intriguing.

On 14th February 1876 two patent applications for a telephone were presented to the clerk in the Washington patent office. The first to arrive was from a lawyer representing Elisha Gray which was put in the in tray of the clerk on duty. The second, a little later in the day, was presented by a lawyer representing Bell. He insisted on the patent being registered immediately, the fifth entry in the Patent Office log book that day, and for him to be issued with a receipt. For this reason Bells patent was deemed to be the first and he gets all the credit whilst

Elisha Gray, whose request was entry thirty nine in the log book gets forgotten.

It was also the case that Bell, being an alien (he was Scottish) had to apply for a full patent with detailed descriptions whilst Elisha Grays application was for a "patent caveat", a less detailed application which was perhaps seen as more speculative.

In the years following there were all sorts of conspiracy theories and a number of court cases but Bell and his lawyers were much better at presenting their case and he kept the credit.

Gray was a successful inventor and with his business partner Enos Barton formed a company Gray Barton which supplied equipment to Western Union Telegraph Company with Gray as Chief Engineer.

He later gave up his position to concentrate on telegraphic inventions, work which was financed by a dentist, Dr. Samuel White of Philadelphia, who had made a fortune producing porcelain teeth. White wanted Gray to focus on the acoustic telegraph (using tones to allow more than one telegraph message to use the same line) which promised huge profits instead of pie in the sky ideas like the telephone.

One consequence of this was Grays undisciplined approach to getting his idea patented at the earliest opportunity as he was not supposed to be working in this area.

Bell was born in 1847 in Scotland, both his father and grandfather were experts in speech and elocution. One of his father's inventions was "Visible Speech", a method by which symbols can represent vocal sounds thus helping people who are deaf pronounce words.

Alexander Melville Bell
Graham Bells father and the inventor of "Visible Speech" some examples of which are shown above

During his childhood Alexander Graham Bell was often co-opted during his fathers lectures. He would leave the lecture room and someone in the audience would be asked to suggest a word in any language which would then be written by his father in visible speech symbols.
Then Alexander would be invited back into the room where, without having heard the original word would pronounce it from the symbols. Over the years he pronounced words in Latin, Gaelic and even Sanskrit. Some examples of the visible speech symbols are shown above.

His father was also the author of *The Standard Elocutionist* which has appeared in 168 British editions and sold over a quarter of a million copies in the United States alone.

Alexander followed in his fathers footsteps and was working as a teacher of the deaf in London and Bath when, in the space of three years, both his brothers died from tuberculosis at the ages of 19 and 25. Understandably worried about their remaining sons health the Bell family decided to move to Canada for the healthier, less polluted, climate.

In 1872 at the age of 25, Alexander was invited to teach at the "School of Vocal Physiology" for the deaf in Boston where his skills with Visible Speech were greatly admired.

Bell (top right) at the Pemberton school for deaf mutes in Boston

He eventually became professor of Vocal Physiology in Boston university, he worked on the problems of teaching deaf mutes to speak and in studying the vibrations caused by speech.

Whilst there he was also paid to teach a five year old boy called Georgie Saunders, whose father owned a leather business and became one of Bells patrons. A second pupil was 15 year old Mabel Hubbard who had become deaf as a baby when she caught scarlet fever. Bell fell head over heels in love with Mabel and eventually married her. Her father Gardiner Green Hubbard also become one of Bells backers but perhaps more importantly was a Patent Attorney, he impressed on Bell the need to document everything he did and to lodge a patent as soon as possible which paid dividends when he beat Elisha Grey to the patent as described above.

Like everyone else Bell had also been working on the methods to use tones to multiplex telegraph messages, this was thought to be the biggest opportunity for riches rather than this outlandish idea of being able to carry speech and this work was encouraged and paid for by Saunders. They had lodged a patent application in 1875 for a Telautograph (a means of sending writing via the telegraph) but the application was not granted as Elisha Gray had already patented something similar a few months before.

Bell's real interest was in the telephone and being an expert on acoustics perhaps approached the problem differently from those who had an electrical or telegraph background; "Had I known more about electricity, and less about sound," he said later, "I would never have invented the telephone."

Although the patent was filed on 14th February 1876 (and granted on 7th March) the first "call" was not made until March 10th when Bell spoke into the device "Mr Watson, come here I want to to see you" and Watson, his assistant, answered. This famous first telephone message which is widely quoted was also the first emergency call, Bell had spilt some sulphuric acid they were using in the prototype transmitter on his clothes which were starting to burn, I am not sure I would use such formal words in the circumstances "Oh **** - help" might have been my first message, perhaps so did he but cleansed the wording when retelling the story.

By June 1876 the apparatus had been improved sufficiently for the first public demonstrations of the Bell telephone. This was followed by Bell showing his new invention at the Centennial Exhibition in Philadelphia in 1876 at which all the latest scientific wonders were displayed most of which were promoted on impressive stands by the large and successful companies. This exhibition followed the Great Exhibition at the Crystal Palace London in 1851 and the Paris Exhibition of 1855. Elsha Gray had a display of his multiple telegraph on the Western Union stand and received a great deal of publicity.

Bell was not very keen to show his new invention yet but Gardiner Green Hubbard put pressure on him to exhibit as did his daughter Mabel who Bell was keen to impress. Bell had little money and his display was tucked into a corner on a couple of tables with a wire strung out between them, no one seemed interested until a remarkable incident.

In the main party of exhibition judges was the Emperor of Brazil, Dom Pedro de Alcantara, as soon as he saw Bell he advanced with both hands outstretched to Bell, and exclaimed: "Professor Bell, I am delighted to see you again."

The judges were suddenly interested in this young inventor who seemed be the friend of an Emperor. Dom Pedro had visited Bell's class of deaf-mutes at Boston University a few months earlier. He was especially interested in the subject and had recently helped to organize the first Brazilian school for deaf-mutes in Rio de Janeiro.

1776 Centennial Exhibition Main Hall (there were many more buildings)

And so, with Dom Pedro in the centre, the fifty judges and scientists crowded round the "first" telephone. While Bell went to the transmitter, Dom Pedro took up the receiver and placed it to his ear. It was a moment of tension, no one had any idea what was about to happen, when the Emperor, with a dramatic gesture, raised his head from the receiver and exclaimed with a look of utter amazement: "MY GOD—IT TALKS!"

Lord Kelvin was also in the party, he was the expert who worked on the first transatlantic cable described in Chapter 2. By now he was the foremost electrical scientist in the world. He listened and nodded his head solemnly as he rose from the receiver and said. "It does speak, it is the most wonderful thing I have seen in America." Elisha Gray was also present and seemed impressed.

You would think that such a glowing testament would immediately result in commercial success or at least sponsorship, it didn't, many scientists just labelled it an interesting scientific toy like the fax machine described earlier.

President Rutherford B. Hayes reflected the views of many when Bell demonstrated the system to him.

> *"That's an amazing invention, but who would ever want to use one of them?"*

Bell had little commercial nous but help was at hand in his future father-in-law Gardiner G Hubbard, who encouraged Bell to promote his invention and with the help of Watson he performed a series of demonstrations. A telegraph wire between New York and Boston was borrowed for half an hour, and in the presence of Lord Kelvin, Bell sent a tune over the two-hundred-and-fifty-mile line. "Can you hear?" he asked the operator at the New York end. "Elegantly," responded the operator. "What tune?" asked Bell. "Yankee Doodle," came the reply.

A series of ten lectures was arranged for Bell, at a hundred dollars a lecture, which was the first money payment he had received for his invention. His opening night was in Salem, before an audience of five hundred people. A pole was set up at the front of the hall, supporting the end of a telegraph wire that ran from Salem to Boston a distance of 16 miles. Watson talked from Boston to various members of the audience. An account of this lecture was sent by telephone to The Boston Globe, which announced the next morning— "This special despatch of the Globe has been transmitted by telephone in the presence of twenty people, who have thus been witnesses to a feat never before attempted—the sending of news over the space of sixteen miles by the human voice."

Bell was an elocutionist and teacher so was good at public speaking as well as having an interesting story to tell so his lectures became very popular. It spread the word about this wonderful new device but that did not result in any meaningful business, however in May 1877 a man called Emery paid $20 for two leased telephones. This was the first time anyone had paid to have a telephone.

Encouraged by this Bell, Hubbard and Watson produced an advertisement promoting the telephone against the telegraph with the messages.

1. Simple to operate, no need for a third person (the morse code operator)
2. Much more rapid; the typical telegraph message speed is 15 – 20 words a minute, the telephone one to two hundred
3. No expense required for either its operation or repair (They were leasing the telephones).

Another early adopter was a burglar alarm supplier called E.T.Holmes who, probably as a publicity stunt, was lent six pairs of phones which he installed in six banks and connected them to the wires he used for his alarms. He put all the paired phones in his office and thus built the first telephone exchange. One of the banks complained and asked for this "toy" to be removed but the others remained for some time. The telephones worked during the day and the lines reverted to being alarms at night.

By September 1877 there were 778 telephones in use and to formalise control of the patents a company was formed called The Bell Telephone Association. Bell, Hubbard and Saunders had 30% each of the shares and Watson 10% but there was no capital – the company had no money. Nor was there any competition, they had a monopoly on the telephone and whilst Saunders and Hubbard were enthusiastic supporters there was an element of helping a friend, Bell had taught Saunder's son to speak for which he was very grateful and Bell was soon to marry Hubbard's daughter. No one else seemed in the slightest bit interested in this "scientific toy".

Alexander Graham Bell　　　　　　　　　　Theodoe Vail
Bell invented the telephone, Vail invented the telephone business

Saunders in particular had invested a large proportion of his money in the venture paying for the workshop, Watson's salary and the manufacture of the telephones, he was close to being bankrupted by it. The company were leasing pairs of phones to a few businesses but it was not paying its way and other investors were not interested. The most likely candidate was the huge telegraph company Western Union and at one point, perhaps in desperation, they were offered the rights to the telephone for $100,000. The company President, William Orton, in rejecting the offer, supposedly said "What use could our company make of the electrical toy?" So he is now known as the man who refused the most valuable business opportunity in history – **It will never catch on.**

He later had to eat his words as a number of reports came into the company of customers replacing their telegraph lines with telephones. Western Unions solution to this irritant? Set up a rival company with the best technical brains they could find (Edison (the inventor), Elisha Gray (the other inventor of the telephone) and Amos Dolbear (another early telephone pioneer who had designed a much improved receiver and the magneto generator for initiating calls) with, in comparison to the Bells funding, the huge seed capital of $300,000.

Their sales pitch announced they owned the rights to the telephone, provided a superior service and prospective customers should not contemplate dealing with the upstart Bell company. Instead of crushing the Bell company it highlighted it's potential to other businessmen. Whilst Saunders and Hubbard were not wealthy they had wealthy friends and by pointing out Western Unions sudden endorsement of the telephone business they raised $50,000 within two months.

The company was now getting off the ground but none of the partners was willing or able to run it. Bell, as we have already seen, was not a businessman, Saunders and

Hubbard were both tied up with their own businesses so there was a need to find a good manager. They appointed a man called Theodore N. Vail who, in his acceptance letter said "My faith in the success of the enterprise is such that I am willing to trust to it, and I have confidence that we shall establish the harmony and co-operation that is essential to the success of an enterprise of this kind." A very flowery 19th century way of saying yes.

He opened a Bell Company office in Reade Street, New York and set to work in establishing the business. Vail had no experience of the telephone business, no one did, but he had the ideal background. His cousin, Alfred Vail, was the man who helped Morse develop the telegraph, perhaps even the inventor of morse code. As a child he had helped his older cousin and became steeped in the telegraph business. He had also worked for the Government Mail Service in Washington and as head of his department had greatly improved the efficiency of the system for carrying post on the railways. In this role he had met Hubbard who was serving on a commission to improve Government mail handling and Hubbard realised his background knowledge of the telegraph and railway networks, together with his proven organisational skills, made him the ideal candidate for this new venture.
It was later said;

> *"Bell invented the telephone, Vail invented the telephone business"*

Vail quickly got to grips with the rather ad-hoc collection of customers and put in place a plan for the business. He also managed to persuade others to back the venture so that within a few months the company had $450,000 of investment and 12,000 telephones in service.

He also stiffened everyone's resolve not to sell out to Western Union, he sent all the agents working in the field a copy of Bells patent with instructions to fight any infringements by others as well as reorganising their contracts so everyone knew where they stood and what territory they were responsible for.

His grand plan, which today seems obvious although at the time seemed an unnecessary complication, was to create a national telephone network with local companies being able to connect to it for national and international calls.

In 1877 an article appeared in the Chicago Tribune which said that Elisha Gray had invented the telephone. Letters were exchanged between the inventors, perhaps with Bell being encouraged by Vail and Hubbard, Elisha Gray replied to Bell in his own handwriting agreeing that Bell was indeed the inventor of the telephone, peace was declared for the time being at least.

One weak link in Bells telephones were the microphones, this was solved by Thomas Edison who developed the carbon microphone. Unfortunately for the Bell Company Edison was working with Western Union who could immediately demonstrate that their telephone was far superior and with a huge network of telegraph wires and a huge financial clout they had a strong sales message.

This setback almost ruined the Bell company, Saunders had lodged his Bell Company share certificates with his bank as security. He received a message from the manager saying "Can you please replace them with a promissory note as the banks inspectors are visiting next week and I don't want to be caught with that stuff in the bank". Salaries were paid late or not at all and Bell, having been to England on a combined honeymoon and sales trip had returned in debt, disheartened and unwell.

Then in 1878 what seemed like a miracle happened. Vail received an offer from an inventor called Francis Blake who had developed a microphone as good as Edisons on the back of work published by Professor David Hughes in England. Blake was keen to exchange the rights to his microphone for stock rather than hard cash. Clearly not only a good inventor but also an astute businessman and ideal for the Bell company as cash is what they didn't have. From a quality of service point of view the new microphone immediately put the Bell Company on an equal footing with Western Union and as a result more investors came forward to put their money in the company, in particular a financier called Colonel Forbes who became company president. By now the company had 22,000 telephones in service, had become the National Bell Telephone Company and had $850,000 of investment.

For several "wilderness" years no one had thought to question Bell's right to call the telephone his invention, after all it was a worthless scientific toy. Now people could see it was perhaps rather more valuable and over the next eleven years there were over 600 patent claims made against him. Some were complete nonsense but others caused a great deal of work for the Bell Company lawyers and expense which started to worry the investors, especially when the mighty Western Union lodged a claim.

In private Western Union had employed Frank Pope, an electrical expert, who spent six months researching hundreds of technical papers, patents and journals and in the end reported to the board that "The only way to make a telephone is Mr Bell's way" and advised them to buy the patents. This advice was not what they wanted to hear so they took the Bell Company to court saying that Elisha Grays telephone, and therefore Western Unions, was the original telephone.

It seemed to everyone apart from Colonel Forbes and Theodore Vail that the Bell Company would be swallowed up by the huge corporation. The case started in the autumn of 1878 and lasted a year during which time the Bell and Western Union fought for every new telephone customer by fair means or foul.

The leading lawyer representing Western Union began to realise they were likely to lose especially when the letter from Elisha Gray to Bell acknowledging him as the inventor was read in court. He advised his clients to settle rather than wait until the end and face the humiliation of being defeated by this small upstart. So, to the surprise of many, Western Union threw in the towel and agreed that:

1. Bell was the original inventor.
2. His patents were valid
3. To give up the telephone business

In return the Bell Company agreed that;

4. That they would buy the existing Western Union telephone system
5. To pay Western Union a 20% royalty on all their existing telephone rentals.
6. To keep out of the telegraph business

At a stroke it added 55,000 telephones in over 50 cities to the Bell network and the Bell Company share price exploded from $50 to over $1000 a share.

With the exception of Vail all the key players now retired from the telephone business having made their millions.

> **Bell** had given all his shares to his wife Mabel. She had a much better grip on financial matters than her husband who was then free to pursue the life of an inventor. He did not like the pressure of being a celebrity nor the continuing hassle of the litigation regarding patents and was keen to go somewhere quiet to work in peace.

When Edison invented the phonograph in 1874 Bell, whose knowledge of sound was second to none, did not think it worked very well so he improved it by using a needle which moved from side to side rather than up and down in Edisons version. It was a great improvement for which Edison had to pay Bell a royalty to incorporate it in his own machine.
Bell's demonstration system had a recording saying "I am a graphophone but my mother was a phonograph". He clearly had a sense of humour.
Eventually the Bells moved to Canada and he did a great deal of work on investigating powered flying

machines as well as becoming President of the National Geographic Society where he coined the phrase "greenhouse gas". He also promoted race relations when he realised how badly his personal assistant (who was black) was treated.

Watson, who as Bells assistant had taken on much of the practical engineering work and in his own name had 60 telephone patents, invested his new wealth in a shipbuilding yard in Boston which eventually employed 4,000 workmen and built ships for the Navy.

Hubbard retired but became very much involved in National Geographic Society.

Saunders invested his money in a gold mine and unfortunately lost most of it.

The other consequence of the Bell Companies sudden rise in value was a plethora of imitators who rushed to market with all sorts of telephone proposals most of which were just scams to con investors into this new boom industry. Just as in the dot com fiasco many dubious claims and valuations were made and an awful lot of money was lost.

Over the years over 600 lawsuits were brought against the Bell Company for patent infringements, none were lost but perhaps if one of the imposters had perhaps looked a bit harder at Meucci's work and supported his claim they might have had a case.

So now the telephone business moved from the "Early Adopter" phase to mass market rollout but there were a few obstacles in the way.

Chapter 6 – Roll-out of the Telephone

A Christmas message from Mark Twain, an early customer of the telephone:

> *"It is my heart-warm and world-embracing Christmas hope and aspiration that all of us . . . may eventually be gathered together in a heaven of everlasting rest and peace and bliss-- except the inventor of the telephone".*

Once the initial euphoria of being able to talk to someone over a distance the popularity of the telephone started to wane. It was expensive and, Mark Twain's biggest complaint, the quality of the calls was poor because:

Single wire plus earth: The early telephones used the same single wire plus earth circuit as the telegraph. Whilst this was fine for a simple on/off connection such as the telegraph it introduced all sorts of noise through the earth connection. This became even more of a problem with the introduction of electric trams and the domestic power grid. In 1882, a former Boston telephone operator, John J. Carty, discovered - as had Bell - that a two-wire circuit (such as is in use today) was considerably quieter but using two-wire circuits doubled the cost.

Crosstalk Even when two wires were used there was interference between adjacent lines if they were run close to each other.

Poor wire: The first telephone circuits were connected using iron or steel wires, these were the same as those used in the telegraph networks. It was always known that copper wire would be much better, however copper was not strong enough to be used in an overhead network.

These issues meant that the telephone failed to sell. By 1880 - when sales should have been growing the novelty of a telephone call was wearing off. While the telephone saved time, was quick and was easy to use it was also expensive, intrusive, and often unreliable.

Many people were content with the telegraph and had got used to the need to write their message down, send it to the telegraph office and wait perhaps an hour for the reply and suggested that the telephone was an unnecessary expense. You have to remember that only a few years before sending a message and receiving a reply on the same day was impossible. Today many people never use the telephone to talk, they just text one another so perhaps 140 years ago people thought the same.

The first breakthrough came in 1877 when Thomas Doolittle developed a process to draw annealed copper wire through a series of dies which greatly increased it's tensile strength making it strong enough to be used as overhead wires.

Up until then the telephone had been used locally but Vail realised the real benefit would be in long distance calling. In 1883 he tried to get the Bell company to run a circuit between Boston and Providence in Rhode Island, a distance of 50 miles, the company were not keen so Vail funded the project himself. At first it used one wire, this did not work and was christened "Vails folly" but then a second was added and it worked well. This proved the concept of two wire working and the company took over Vail's line and ran a two wire copper circuit from New York to Boston, a distance of 220 miles at an astronomical cost of $70,000 ($1.8m today). As soon as it went into service it proved an instant success but to replicate this required the Bell company to rewire the whole of the USA!

The companies could only make money if they managed to attract a lot of customers but now the engineering problems were solved the long distance market started to take off. Being able to talk to someone in the same town or city was useful but to be able to communicate with someone 50 or 60 miles away at a moments notice was a real benefit. The alternatives being the telegraph which meant going to the local telegraph office, dictating a message, it being decoded at the other end and a messenger delivering the message to the recipient did not allow a practical "conversation" nor did it allow privacy or discussion of an issue. And of course travel was more difficult than today, you could not hop in your car and go and see someone for a meeting.

The good business case for long distance telephony encouraged investment in higher quality cables between cities. The thicker the copper wires the lower the resistance and the further the network can reach with 1000 miles being possible. The callers however had to visit the telephone company's offices at either end to use "high powered" telephones. At this stage it was not possible to connect the local network into these very long distance lines.

One downside of using copper was that some members of the local Indian tribe greatly admired the uninsulated shiny copper, which glowed in the sun, and they stole the lines to make jewellery temporarily interrupting long-distance service. Replacement copper lines were dipped in tar; these new lines remained intact.

The long distance market became profitable especially as the calls could be timed (by the operator) ensuring a cost effective use of the cables. This contrasted with the local network where customers could stay on the lines for as long as they wanted. AT&T was spun off as a separate company to manage the lucrative long distance business. It was not until the invention of the valve in 1907 that very long distance communication became possible. Valve repeaters were first used in the UK network in 1923 greatly improving the quality of long distance calls and reducing the amount of copper needed in the long distance cables.

In cities the mass of telephone wires became a nightmare and in 1882 Theodore Vail set up a number of trials to find the best method of running cables underground. They used gutta-percha for the outer covering and paper as insulation between wires and from then on telephones in cities were connected with cables buried underground. The problem of crosstalk, where one conversation could be heard on another line was solved by twisting the pair of wires in the same circuit together so that one wire counteracted the signal in the other.

A major improvement came in 1888 when Vail sponsored an engineer called John Barrett who worked out how to sheath the cables in lead. This provided a flexible, waterproof covering which could be jointed using a blow torch. Lead was used right up to the 1950's after which polythene was introduced but lead covered cables were still common when I was an apprentice in the 1970's and the technicians who could plumb the joints were justifiably proud of their skills. I was given a bit of scrap cable to have a go at plumbing when I was a trainee and ended up with a burnt cable (the insulation was paper don't forget) and a big blob of lead on the floor of the manhole much to the amusement of the men I was working with.

One side effect of the initial use of overhead wires was that the first switchboards were always on the top floor of buildings as that is where the wires came in.

In one attempt to generate interest concerts were "broadcast" over the telephone so people could hear them without leaving home. This was long before radio was invented.

Meanwhile in the local Bell Company in the city of Rochester tried to introduce charging for local calls, this backfired horribly. A journalist called the new rates "the final culmination of inordinate greed," and nearly every prominent citizen in the city stopped using their phones. When the Rochester Council threatened to cut down the ninety-foot telephone poles on Main Street the company relented and returned to flat-rate service. The boycott ended after eighteen months.

Forbes, the Bell Company president, and Vail, the General Manager, fell out in 1888. Forbes was only interested in returning money to the shareholders and with profits doubling every year was popular with investors, Vail on the other hand wanted to reinvest in expanding the network and improving service especially as he could see the fast approaching time bomb of Bell's patents expiring. Forbes won and Vail resigned.

Bell's patents expired in 1893 and 1894, after which many other operators entered the market and set up local networks serving individual towns or niche markets such as farmers who often ran in their own wires across their land. AT&T refused permission for any of these rival companies to connect to their long distance network so other long distance rivals set up in competition. This meant that anyone who wanted to call a variety of people needed to subscribe to more than one network. Not a great situation and it made running a telephone company an unprofitable business.

> *One of the very most useful of all inventions, but rendered almost worthless & a cold & deliberate theft & swindle by the black scoundrelism & selfishness of the companies of chartered robbers who conduct it.*
>
> Mark Twain

Twain was clearly not happy with the service and nor were many other customers. The Bell Company was soon in financial trouble but fortunately Vail was enticed to return in 1907. He brought about a rationalisation and eventually just AT&T controlled the national network in the USA and each local area had one telephone company who could then connect to the AT&T network.

Theodore Vail was held in such high esteem that when he died in 1920 all the telephone companies in the USA paid tribute, the entire telephone service of 12 million customers was shut down for a minutes silence.

Theodore Vail in 1918

So what was happening in the UK?

On 14th January 1878 Bell had demonstrated the telephone to Queen Victoria at Osborne House on the Isle of Wight with calls to Cowes, Southampton and more impressively, London. He was actually on his honeymoon at the time, the records don't say whether he took his wife to meet the Queen. These were the first long-distance calls in the UK and the first to be connected under the sea. The Queen was impressed and she became an early adopter, her endorsement gave the marketing of the telephone in the UK a significant boost.

This helped promote the Telephone Company of London who had the rights to Bell's patents in Great Britain. The company was registered on the 14th June with a capital of £100,000. It had a capacity for 150 lines and opened with 7 or 8 customers. Two more exchanges were soon added in central London as well as exchanges in Glasgow, Manchester, Liverpool, Sheffield, Edinburgh, Birmingham and Bristol.

As in the USA where the Bell and Western Union companies were aggressively chasing prospective customers a rival was soon in the market, the Edison Telephone Company with the Edison/Western Union telephone patents. Their first exchange officially opened on 6th September 1879 in Queen Victoria Street, London, with ten subscribers who used the superior Edison carbon transmitters. By the end of the following February, when the company had another two exchanges in operation, it served 172 subscribers. The annual tariff was £12 against £20 charged by the Bell Company.

Mr. William Preece (later Sir William Preece) of the Post Office Engineering staff, when asked whether the telephone would be popular with the public, replied

"I think not, I fancy the descriptions we get of its use in America are a little exaggerated; but there are conditions in America which necessitate the use of instruments of this kind more than here. Here we have a superabundance of messengers, errand boys, and things of that kind."

So clearly in 1879 the Post Office establishment did not think of the telephone as a worthwhile business, no different to Western Union – "**It will never catch on**"

A year later the Government in the form of the Post Office finally took an interest in the telephone and decided to regulate the market. In a famous High Court judgement in December 1880 the court ruled that the telephone was really a form of telegraph and as such was subject to the telegraph rules and regulations including the obligation to obtain a licence and pay the Post Office 10% of gross income. The Post Office was also given the option to purchase the business when the licences came up for renewal.

At this point the Post Office themselves started to offer a telephone service and build exchanges.
It soon became clear to both the Bell and Edison Companies that the new licencing rules together with Western Union losing the case against Bell in America that arguing over patents and setting up costly rival networks was not a great business plan so they merged to become the National Telephone Company Ltd with the rights to all their combined patents.

The situation was then thrown into some confusion when on July 1882 the Postmaster-General, Henry Fawcett, decided to grant licences to operate telephone systems to all responsible persons who applied for them, even where a Post Office system was established, reversing the previous policy of the Post Office being the main provider 'on the ground that it would not be in the interest of the public to create a monopoly in relation to the supply of telephonic communication'.

Two years later he also withdrew a restriction which limited licence areas to five miles. Instead, telephone companies were allowed to work anywhere in the United Kingdom and were thus able to create exchange areas of any size and to connect them by trunk wires. This opened the way for the development of a national trunk system. This relaxation of the rules by the Postmaster-General also encouraged the introduction of the public call office. Telephone companies were now allowed to establish telephone stations which any member of the public could use. There were little more than 13,000 telephones in use at this time and the decision allowed access to the telephone to a whole new sector of society. The new 'call offices' were soon advertised in the national and local press. They were at first located in 'silence cabinets' found in shops, railway stations and other public places as described in Chapter 8.

In March 1892 the Postmaster-General had a new plan largely as a consequence of complaints over the quality of the National Telephone Company's service and the accumulation of its overhead wires in towns. He announced the Government's proposal to purchase the trunk lines of the National Telephone Company, telephone companies would also be restricted to local areas under new licences. The Post Office was also very concerned about the increasing competition of the telephone which was now eating into the revenue from the telegraph services. The new policy was brought into law by the Telegraph Act which made provision for the raising the money for the purchase and extension of the trunk telephone system.

It took a few years to sort out the detailed arrangements but in 1896 an agreement was signed with the Government paying £459,114.3s.7d for the 29,000 miles of the National Telephone Companies trunk cable (equivalent to £67 million today). Under the agreement Post Office customers could make calls via the trunk network to other companies customers or vice versa but there was no provision for interconnection at a local level. So if you were a customer of the National Telephone Company in Glasgow you could not call Post Office customers who also happened to live in the city.

The government were also keen for local authorities in provincial towns and cities to set up local networks and in 1899 an act was passed to encourage this. Fifty five Councils applied for licences but only six actually started a service and in the end only one survived, Kingston on Hull who remained part of Hull Corporation. They were one of those British anomalies which seem to feature in pub quizzes, the most striking difference to the rest of the UK was that their public phone kiosks were cream, not red. That changed with the telecoms liberalisation in which Hull Council benefited from a windfall when it's telephone business was floated as a public company called KCOM in 1999.

The first telephone users called the exchange by turning a handle on a magneto generator and had local batteries at their premises providing current for speech. Magneto generators were expensive and the local batteries, which had to be kept near the telephone, were bulky and prone to faults. When we complain of having a low battery in our mobiles we think it a new curse, it isn't phone users in 1880 had the same problem.

These limitations were removed when in 1882 G.L.Anders of London patented a central battery system by which telephones could be supplied with electrical power from the exchange without the side effect of every line being connected to every other one.

If your home telephone is connected via copper (over 95% of the UK) then the power (48v DC) is supplied from batteries in the exchange. Until the 1980's these were huge lead acid batteries with each cell the size of a hot tub with 24 making up one battery.

For security there was always more than one battery and a technician checked the specific gravity each day to make sure they were fully charged. Charging was from the mains supply or a generator if this failed. The batteries were designed to provide enough power for 24 hours giving ample chance for portable generators to be brought in on lorries in the event of a major failure. In some large city centre exchanges that battery room was enormous with many thousands of gallons of sulphuric acid gently bubbling away. Just like a laptop which is plugged into the mains continuously the battery just serves as protection against surges and a backup in case of a power cut.

To get a discount on their electricity bill some American operators agree to allow the power companies to turn off their supply in peak periods and to run on generators for a few hours. In a major New York exchange in the 1980's this plan was instigated but for some reason the generators did not start and no one noticed until the battery went flat and the exchange failed. It took days to restore service!

To generate more revenue streams and encourage the take up of the telephone in 1884 the "Electrophone" service was launched in London. The Company "broadcast" news and entertainment over the telephone, just like a radio studio today they read news reports connected the service and had live feeds from theatres or concert halls.

The service cost £5 a year and the customers were issued with headphones so they could sit back and relax whilst listening to the programmes, this was 10 years before radio was invented and at least 40 years before wireless programmes were first broadcast. Customers were also issued with a special "answer back" microphone so they could talk to the Central Office and request different programs.

By the end of the first year, Electrophone had forty seven customers which increased to around 600 by 1908 and 2000 at it's peak in 1923. The company covered performances from some 30 churches and theatres. In 1923 an Electrophone director was quoted as saying that "it would be a long time before broadcasting by wireless of entertainments and church services attained the degree of perfection now achieved by the electrophone" what he meant was the BBC who started transmitting in 1923, **it will never catch on**!

But Electrophone could not compete with radio and the service ended in 1925. Annual subscription service for music? – Who said Spotify was a new idea?.

Another service which has lasted a little longer is TIM, the speaking clock, launched in 1936.

The novelist Conan Doyle gave his detective Sherlock Holmes a telephone at his home, 221b Baker Street but in practice in the stories almost all his communication was by telegram, none of the people he wanted to contact would have had a telephone.

Exchange Battery Room

Electrophone Advert

Chapter 7 – The Automatic Telephone Exchange

The first telephones were bought by people who wanted quick verbal communication between offices or home and office. Understandably this was a limited market and fairly soon the switchboard was introduced to allow any customer to talk to any other *on the same network*. This is an important caveat, in many cases in the early years customers on one telephone network were unable to call those on another. When the UK monopoly was broken in the 1980's with the introduction of Mercury and the mobile operators interworking, as it is known, was still a very thorny issue but more of that later.

The first time several telephone lines were routed to one point was when the burglar alarm company run by Edwin Holmes put six phones connected to six different banks on the same shelf as described in Chapter 5. In those very early days the caller shouted into the microphone which alerted the person at the other end.

The first proper switchboard was designed by Tivada Puskás a Hungarian inventor living in America who was working on his idea for a telegraph exchange when the telephone was invented so he modified his designs to manage this new technology. The first experimental telephone exchange was based on the ideas of Puskás, and it was built by the Bell Telephone Company in Boston in 1877.

The basic idea of a switchboard was simple, the customer wound a small handle on their telephone which generated an AC signal which in turn activated an alarm on the switchboard. The operator then plugged a cord into a socket associated with the callers line and spoke to the caller, was given the name of the person they wanted to speak to (numbers came later). Another jack was then plugged into the called persons line. First the operator sent an AC signal generated by their own hand cranked magneto to ring the bell, then, when the person answered, the caller and called customers lines were connected together. The operator regularly listened into the call to determine when it was over and they could then unplug the connections.

Sounds simple. Well it was at first with only a few customers but as the networks grew so did the size of the switchboards. One problem was that the operator had to manage the cords and switches as well as noting down a record of the call (for billing trunk calls) so needed both hands free. Step forward French engineer Ernest Mercadier, who was awarded a patent for the first ever headphones in 1891.

As the switchboards and networks grew the operators needed to pass calls between each other both in the same room and to other towns and cities. Routing a call became ever more time consuming and complex, something the paying customer probably did not appreciate. To simplify the process the telephone number was invented so the problem of several Smiths getting wrong calls was removed.

Emma Nutt (centre) the first female operator

At least by the 1960's the operators were sitting down

Manual switchboards remained largely unchanged for almost 80 years. I can remember talking to someone who was an operator in a manual exchange on 22nd November 1963 who told me suddenly the whole place lit up with everyone making calls at the same time why? President Kennedy had been shot and people wanted to share the news with their friends. These sudden peaks of telephone traffic are quite difficult to deal with even now with electronic exchanges. In the past with only a limited number of operators it was a real issue and at times made being an operator quite a stressful job with irate customers having to wait a long time.

During the Wall Street panic in 1907 one New York switchboard handled 15,000 calls in an hour, it must have been quite hectic.

At first the operators were boys or young men, the work required them to stand, kneel and stretch, boys were nimble, quick and they were also cheap to hire.

So when Bell Telephone launched its first phone systems, in 1878, the company hired teenage boys as its operators but they used to play pranks, drink and be rude to customers and their customer care skills were not good.

This was solved when women were employed instead, they had a softer approach to the customers and were more careful with their language.

The first female operator was Emma Nutt who was appointed in 1878. Manual exchanges remained in service in the UK for almost 100 years, the last in London (Upminster) was finally replaced in 1970. Incidentally, the last in America was not automated until 1982 by which time it had become something of a tourist attraction.

Switchboards remained in service in the UK for another ten years in order to route national and international calls.

When I started work as an apprentice in 1969 the local exchange had about 20 engineers maintaining the equipment and 100 operators manning the switchboard which dealt with trunk, international calls and directory enquiries.

It was an awful job, the girls (and it was all girls during the day and then men overnight (11pm to 8am)) were treated very badly. They sat at their position being watched by a supervisor, invariably a dragon, who seemed to be appointed by being the longest serving operator not the one with the best management skills. The girls had to clock in and out, not just at the start and end of the day but for tea breaks and lunch as well. If the operator wanted to go to the toilet she had to put her hand up and, if permitted, could leave her position to visit the toilet. Her trip was timed!

In the close knit environment, chained to their positions, the girls were always looking for a bit of amusement, every apprentice was sent to work in the switchboard room at some point and as a spotty 16 year old you were subject to all sorts of teasing.

In 1889 a Kansas undertaker called Almon B Strowger became aware that when callers asked the operator for an "Undertaker" they were always connected through to his rival whose wife worked as an operator, he also suspected the other operators received a commission on calls connected. We complain about Google pushing particular products when we make a search, this is nothing new the operators were perhaps the original Google!

To overcome this he invented the automatic telephone exchange. In solving the problem he first had to invent a device to send pulses to indicate each digit between 1 and 10 (0) as well as a switch which could select the right circuit based on those pulses.

In his first version the caller used two telegraph type keys on the telephone, which had to be tapped the correct number of times to step the switch, but the use of separate keys with separate conductors to the exchange was not practical for a commercial system. He soon developed a dial which made the process of sending pulses an awful lot easier.

Incidentally when I was a child it was possible to tap out the digits on the handset cradle of a payphone and make local calls for free.
Early advertising called the new invention the "girl-less, cuss-less, out-of-order-less, wait-less telephone"

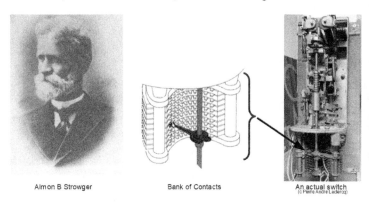

Almon B Strowger   Bank of Contacts   An actual switch
(c Pierre Andre Lederoq)

The Strowger Automatic Telephone Exchange Company was founded in 1891 and the worlds first public automatic telephone exchange, using Strowger's automatic telephone system, was installed at La Porte, Indiana (Strowgers home town) in November 1892; 45 subscribers were connected.

The Strowger switching system proved extremely popular and in 1922 was adopted as the standard for all automatic telephone exchanges in the UK. This electro-mechanical technology persisted for over seventy years. The UK network of over 6,700 telephone exchanges were mostly Strowger based technology. These were gradually replaced by more modern exchanges culminating in the closure of the last working Strowger electro-mechanical exchange at Crawford, Scotland on 23 June 1995.

There are many videos on uTube showing strowger exchanges in operation (https://www.youtube.com/watch?v=HcvA5q8yOTo is a good example)

In 1911 The Automatic Telephone Manufacturing Company Ltd of, Edge Lane, Liverpool, was formed to exploit the UK Strowger patent rights of the Automatic Electric Company of Chicago, the proprietors of the patents. It was the first manufacturer of automatic telephone equipment in the UK The Edge Lane factory became Plessey and finally GPT before closing in 2005.

One limitation of the telephone network was the need to keep the line between the exchange and the customer to less than three miles so that the resistance of the wire is kept within reason (with newer technology this was extended a little from the 1960's but then broadband reduced it again). So the UK telephone network needed between 6,000 and 7,000 exchanges to be able to work within these limits. This in turn meant that many exchanges only managed a few customers and were therefore served by small manual exchanges attended by caretaker operators. Exchanges with fewer than 20 subscribers did not normally give service at night or on Sundays, an obvious inconvenience. The first rural automatic exchange in the UK was brought into service in October 1921 at Ramsey, Huntingdonshire, it had 40 lines and overcame the problem of providing a 24 hour service.

Anyone who has ever been in a large Strowger exchange

will not have forgotten it, the noise is deafening! The mechanism used to move the contacts uses electro magnets which slam a ratchet. It is hard to describe but you can see it in action in many utube videos.

Amazingly the UK was still building Strowger exchange equipment in the 1970's. One suggestion was that as the equipment was manufactured in the constituency of Prime Minister Harold Wilson that he put pressure on the Post Office, a government owned monopoly, to keep the production lines running.

In the 1950's a more reliable (and quieter) exchange design called "crossbar" was introduced in Europe and the USA. It consisted of a matrix of relay contacts which could route the calls. It worked reliably, had fewer moving parts and was therefore cheaper to maintain.

The British Post Office decided to keep on with Strowger whilst pinning its hopes on electronic exchanges. The ultimate dream was to build the first digital exchanges and in 1962 the first truly electronic exchange went into service in Highgate Wood in London with 800 lines. It was a joint project between the Post Office and all the UK manufacturers with the intention of producing a world beating design.

You have probably heard of Alan Turing of Enigma fame, he was the mathematical genius who worked out the logic to decode the German messages during the war and designed the first computer. He only did half the job, someone had to actually work out how to build an electronic computer and that man was Tommy Flowers who was on loan from the Post Office research department at Dollis Hill.

Tommy designed and built the first Enigma decoders out of bits of Strowger exchanges and then, when a more complex machine was required, designed and built the first electronic computer called Colossus using 1800 valves. It was so successful that several were built before the war came to an end.

The doubters said that a machine with so many valves would not work as at any one time there would be several that had failed.

What Tommy showed was that, unlike radio sets or other devices with valves which are turned on and off, if you leave the equipment switched on and running it is very reliable.

 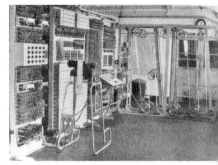

Tommy Flowers — His Colossus computer built from telephone exchange equipment

After the war Tommy went back to the Post Office as the head of telecommunications research and was driving force behind the Highgate Wood exchange. Because of the secrecy around the Enigma decoding work he was unable to prove it was possible to build reliable valve based computers and in the end the exchange was shut down and the experiment regarded as a failure.

It was not until the mid 70's, 30 years later, that the story broke about the Bletchley Park work. It is hard to imagine a secret staying secret for as long nowadays and Tommy and for that matter Alan Turing, did not get the credit they deserved.

There is a lovely story that in 1993, aged 87 Tommy received a certificate from Hendon College, having completed a basic course in information processing on a PC. He had designed and built the first electronic computer before Bill Gates and Steve Jobs were born!
As this experimental exchange was not a success the Post Office, again in cohorts with the manufacturers, designed a series of exchanges based around reed relays. Reed relays consist of two gold plated metal strips in a glass tube inside an electro magnet. When the magnet is

activated the strips became small magnets and make contact.

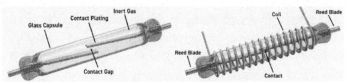

The advantage is that, being hermetically sealed, no dirt could get in and they would be reliable which in the most part they were (but see below). The control was electronic, transistors, no valves and integrated circuits had yet to be invented. The first version aimed at rural exchanges of up to 2000 lines was called TXE2 (Telephone eXchange Electronic). About 3,000 of these small exchanges were built and a few were exported, mostly to friendly telecoms companies such as Gibraltar Telecom

Its big brother TXE4 was designed to handle 40,000 lines, being designed by manufacturers committee meant there were lots of teething troubles such that the first exchanges did not go into service until the mid 1970's by which time digital technology was on the horizon, about 100 TXE4 exchanges were installed, none were exported.

Each TXE4 had hundreds of thousands of reed inserts which were supposed to be very reliable assuming that the manufactures quality controls were good. In the early days several batches started to fail and a few exchanges had to be re-reeded, a costly and time consuming process. The rumour at the time was that the staff on the factory floor were gold plating their own objects in the electroplating vats used to coat the reeds, including cigarette lighters the fuel of which ended up in the gold and therefore in the reed relays!

The real breakthrough came with the worlds first digital exchanges which were much more reliable and cost effective. What do we mean by digital? All is explained under PCM in Chapter 11.

The first digital exchanges went into service from 1972 and were made by Alcatel of France. BT's System X first went into service in 1980 and in the USA the 5ESS from Lucent was introduced in 1982. System X was another joint BT / Plessey / GEC development which was aimed at the requirements of the UK network which made it difficult for other manufacturers to compete in the UK but this uniqueness of System X was also a great handicap when Plessey and GEC tried to sell it overseas.

To introduce a bit of competition in the supply of exchanges BT ran a competitive tender in the early 1980's for "System Y". The tender was won by Ericsson who established a large support organisation near Brighton. To adapt their AXE10 digital exchanges to meet the UK quirks in the services provided to customers Ericsson had a significant amount of development work to do. Their first exchange went into service in 1986.

There was a little bit of a problem with the Ericsson name. Ericsson of Sweden had started manufacturing in Nottingham in 1903 but during WWII the government were concerned that Sweden was too closely aligned with Germany (they were officially neutral) so Ericssons shareholding was reduced to less than 50% and the Swedish Managers returned home. From then on the company was under British management and in 1961 became part of the Plessey Company but they still used the Ericsson name. So when Ericsson won the System Y contract they had to ask Plessey (who made System X) for their name back so they were able to trade in the UK.

By 1998 the whole UK network was digital, the next stage was to move to an IP (Internet Protocol) based network, this has taken rather longer than at first thought and 30 years on many of the System X and AXE10 exchanges are still in service.

There are two issues facing any new telecoms technology;

1. Interworking with several generations of old equipment
Just because the telephone company is upgrading its exchange should not mean every customer has to change their phones or equipment. There are all sorts of old legacy systems connected to the network, phones with rotary dials, fire or burglar alarms, fax machines etc as well as other telecoms companies who do not have the latest version of signalling. I suspect somewhere in the world there is still a Strowger exchange which a caller in the UK or America might wish to call. You cannot just replace the existing systems with something new, this is less complex than it used to be but it is still an issue.

2. Reliability
I can remember working on some joint venture bids on behalf of Ericsson with a number of computer vendors who always looked down their noses at us "Old Proprietary Technology" companies but they always went a bit vague when downtime was mentioned. They always said their equipment has a 99.999% reliability or some

similar figure. 0.001% of a year is 8.7 hours, BT used to insist the maximum downtime per year of an exchange was 2 minutes! This includes periods due to software upgrades. I cannot say this level of service was always met but if an exchange was out of service for more than 5 minutes in a year then there would be a large scale enquiry as it would be an unusual event.

It was suggested that the exchanges should be managed by a man and a dog. The man's job was to feed the dog, the dog's job was to bite the man if he went anywhere near the equipment. This was brought home to everyone in 1987 when the BT engineers went on strike. With Strowger it would only have been a matter of days before things would start to go wrong but the digital exchanges just kept going, the engineers went back to work after four weeks. It proved if you don't touch the equipment it does not go wrong

BT aim to replace all the System X and AXE10 exchanges by 2025 with their IP based network on which they can manage all voice calls, both fixed and mobile as well as data. They would like to move everybody onto a fibre connection rather than copper. The customer will then have almost no restriction of the speed of their broadband service although I guess there will then be charges relating to how much you use. You will also need to provide power, at the moment the exchange batteries provide the 50v which powers your land line phone which means it still works in a power cut. How will BT move the customers who just use a phone remains to be seen?

Chapter 8 – Phone Kiosks

The 'liberalisation' by the Postmaster-General in 1884 brought about the birth of the public telephone. At the time there were only 13,000 telephones in the UK so the decision allowed access to the ordinary man or woman in the street for whom the telephone was largely only a rumour was revolutionary. The public telephones were at first located in 'silence cabinets' found in shops, railway stations and other public places.

The first free standing call office (later to be known as 'kiosks') was introduced in Bristol in 1886. It was basically a small wooden hut where a three-minute call could be made for two pence (a little under 1p). Not all early payphones had a coinbox built into them; some of the kiosks had a penny-in-the-slot mechanism on the door, while others had an attendant to collect the fee.

The National Telephone Company provided subscribers with a key which were used to unlock call offices when they wished to make a call in the attendant's absence. Unless you knew the person you wanted to call would be standing next to another kiosk (and you knew the number) you could only call people with their own telephone. This was likely to be doctors, hotels, shops or businesses.

The Post Office's first coin-operated call box was installed by the Western Electric Company at Ludgate Circus, London in 1906

The first Post Office standard kiosk design was similar to the old wooden-box version, but was made up from three sections of reinforced concrete and fitted with a wooden door with the two sides and front containing glass panels. Once the kiosk had been constructed it could was then painted any colour to meet local conditions. It had the motto "Open Always" on the glass panels emphasising the 24 hour nature of the telephone. An initial contract had been placed with Somerville & Company in March 1920 for the supply of 50 kiosks at a price of £35 each.

Clearly the Post Office did not envisage a huge demand but in time they did prove popular and eventually over 6,000 were made.

In 1924 a competition to design a new kiosk was organised and several leading architects were invited to submit designs. Models were placed on view behind the National Gallery in London and selection was made by the Fine Arts Commission. The winner was a design by Sir Giles Gilbert Scott and, after a slight modification to the door and change of material from mild steel to cast iron, it was adopted by the Post Office and designated Kiosk No. 2, or K2.

Some important improvements to the door mechanism and window arrangement were contained in the kiosk. The glass was deliberately made into small panels so that breakages could be repaired with a minimum of renewal. There was also a ventilation system which worked through perforations in the dome. Because of its cast iron construction it weighed approximately 1.5 tons and had more interior space than its predecessor. The most distinctive feature was undoubtedly the bright red colour scheme. The kiosk's introduction in 1927 was mainly confined to London and some large provincial towns and proved to be very successful. Over the next ten years there were a few improvements (models K3, K4 & K5) but in 1935 the K6 became the standard kiosk of which 60,000 were made and is what became the iconic red telephone box although to the untrained eye it does not look very different to the original Gilbert Scott design.

Gilbert Scott's original model of what was to become the K2 stands at the entrance of Burlington House, the home of the Royal Academy in London.

In 1925 new type of coin-box mechanism was introduced, known as the Button A and Button B prepayment equipment, and for over 25 years its design remained unchanged despite various developments in the design of kiosks.

To make a call users inserted the appropriate fee which prepared the circuit for dialling, when the call was answered the caller then pressed Button A. This allowed the coins to be deposited into the cash box and the call to be connected. If a call could not be connected for some reason, or if there was no reply, Button B was pressed and all the coins were returned to the caller. There were all sorts of problems over the years as inflation increased the cost of calls the boxes had to be modified to accept different combinations of coins.

In 1959 the first versions of the new Pay-on-Answer payphones were being introduced and at the end of the 1950s began to supersede the 'Button A and B' models in order to allow trunk calls using Subscriber Trunk Dialling (STD) rather than the operator.

Silence Cabinet

Kiosk Number 1 (K1)
Avoncroft Museum

Coin Box
Press A to speak and B for your money back

When Mercury started to compete with BT they installed their own phone kiosks, particularly in prominent locations such as railway stations and at busy locations in major cities. Running telephone kiosks is not a profitable business even before the advent of mobile phones which make them almost redundant. They need emptying of cash on a regular basis, are liable to vandalism and being outside are more prone to faults, however they did raise Mercury's profile as the boxes were quite distinctive but after less than 10 years they sold that part of the business. We tend to forget how vital they used to be, I help with a local charity walk which has just celebrated it's 25th anniversary, the organiser explained that the various checkpoints over the 24 miles were not located where it was most convenient for walkers but they had to be sited next to phone boxes as 25 years ago no one had a mobile phone!

The Post Office were not the only providers of telephone boxes, the AA and RAC also put telephone boxes in busy locations and gave their members a key which enabled them to get into the box and call the helpline in the event of a breakdown. The Police also had boxes which anyone could use to call 999, the most famous of these boxes is the Tardis used by Dr Who. Only about 1,000 Police Boxes were installed, the last closed in 1981.

Chapter 9 – Telephone Instruments

I don't want to spend a great deal of time describing all the different types of telephone as that would make for very boring reading. All telephone instruments have the same basic functions:

1. To convert speech into a variable current
2. To convert a varying current into sound.
3. To indicate to the exchange that the customer wants to make a call
4. To ring a bell or some alarm to indicate an incoming call.
5. To be able to signal the telephone number required.

The first telephones had a funnel like microphone and a separate receiver which the customer held to their ear. Most were wall mounted and the caller stood in front of them talking into the microphone and holding the earpiece.

The unit, invariably made of nice polished wood, was an expensive luxury so it had to look the part, it also had a bell and a small handle which the caller rotated to call the operator. This handle generated an alternating current which was sent down the line to ring a bell on the switchboard. Before 1900 the unit also needed to house the batteries needed to power the telephone.

People wanted a much smaller device which ideally they could have on their desks so the Candlestick telephone became popular with the earpiece and microphone on a small stand of the type you often see in films and pictures of the time. What you rarely see is the associated box which holds the bell, capacitor, batteries and the magneto used to call the exchange.

By 1920 the need for the magneto and batteries had gone but, where there were automatic exchanges, a dial was required.

The dial was clockwork, you put your finger in the hole displaying a number between 1 and 0 (ten) and wound it clockwise to an end stop. By letting it go the dial returned to rest and at the same time broke the connection to the exchange by the same number of times as the digit required.

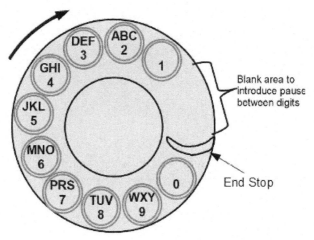

So if you dialled 8 the eight short breaks were made which moved the Strowger selector eight steps. There was a blank section of the dial, this ensured a slight delay so the subsequent digits did not run into each other, so if you dialled 2 followed by a 3, there would be two breaks, a gap followed by three breaks.

Before the 1950's exchanges were given names and to call a number on that exchange a three letter code was used. A famous number was Scotland Yard whose number was

Whitehall 1212 so the caller dialled WHI 1212 (9 for W, 4 for H and 4 again for I then 1212).

The first handsets appeared from around 1910, that is the microphone and receiver in the same hand held unit and that basic design remained unchanged until cordless phones were introduced in the 1980's.

Until the 1980's the Post Office / BT owned the telephone instruments which they rented to the customers. If they went wrong then a man (there were no women engineers until the late 70's) would come round to fix it. The most common problems were the dials running fast or slow or the wires between the handset and the main unit failing. Push button phones were introduced in the 70's but they still converted the button pressed into break pulses to satisfy the Strowger exchanges.

Even when TXE and crossbar exchanges came into service BT seemed reluctant to change from this old technology whilst the rest of the world moved onto tone signalling, the different buttons generated a different tone which was decoded by the exchange, it made dialling a number a great deal faster but if the exchanges themselves were old fashioned Strowger (which most were) the caller had a long wait before hearing ringing tone.

Whilst they supplied the phone BT were also very reluctant to change the design, the only attempt at modernity was the Trimphone (Tone Ring Illumination Model) which style conscious customers could have at a higher rental from 1968. They might be style conscious but they also took a risk, the luminous dial was mildly radioactive, the Atomic Energy Authority in Harwell were find £3000 when a skip was found at their site with several thousand trimphone dials in it exceeded the legal radiation limits.

Standard Phone
(called a 706)

Trimphone
(called a 722)

All this changed when BT was privatised and a white socket was installed in every customers premises to which they could connect any approved phone bought on the High Street. This had quite a big impact on the manpower requirements, no longer were engineers visiting peoples homes to mend the telephones and doubtless enjoying a cup of tea, BT's responsibility ended at the little white box on the window sill.

Chapter 10 – Telephone Numbers

Telephone numbering might seem a boring subject but it is a vital one.

We are so used to picking up the phone and dialling a code with which we can directly contact anyone else in the world. There is quite a bit of technology involved to manage this. If you are making a call from Oxford to Cambridge you dial 01233 followed by the local number, if you were calling from Aberdeen the code is still 01233. Oxford to Cambridge may have only required one intermediate exchange but many more hops for Aberdeen to Cambridge.

For this reason until the 1950's and 60's long distance calls were made via an operator who understood the network topology, the customer just asked for a call to Cambridge 223344.

The use of operators was very expensive and could become a bottle neck as if there was a sudden surge in demand and there were not enough operators on duty.

To overcome these limitations (and costs) an electro mechanical system was developed which allowed the caller to dial a universal code (i.e. 01233 for Cambridge) no matter where they were in the country and the subscriber trunk dialling equipment would translate this into the digits necessary to route the call from that location. This meant there was sometimes quite a delay between dialling the number and getting ring tone but at least you did not have to wait for an operator.

The Subscriber Trunk Dialling or STD system was introduced from 1958 often with a high profile personality asked to make the "first call". The first was made by the Queen who dialled a number in Edinburgh from Bristol.

The Post Office were always terrified about these calls going astray so inevitably had some sort of emergency plan in place. In the 60's a local mayor in London was asked to initiate the first STD call from his area, the press were invited and there was a special telephone set up for the purpose. The idea was that the mayor would call the mayor of Newcastle who was presumably standing by for this call. Just in case the Post Office set up a circuit between the two locations and a technician was told to monitor the call, if it looked like it was going wrong to switch in the backup circuit.

The mayor arrived a little early and asked the Post Office big-wig attending to him if, while they were waiting, he could call his wife, "of course" was the answer. He picked up the special phone, dialled his local number but miraculously ended up speaking, not to his wife but to the mayor of Newcastle thanks to the engineer who saw that something was going wrong.

But getting the caller to dial right the number is not fool proof. I remember a case where we in BT were in serious trouble with a business customer who was getting hundreds of mis-routed calls from people who had been cut off from their Satellite TV service. The company were understandably annoyed at their receptionist spending all day answering wrong numbers.

Many hundreds of man hours were spent checking all the data in the exchanges as well as test calls from all over the country to try and establish where the fault was. There was no fault; the satellite company's letter, telling the customer they had been or were about to be, cut off had two numbers at the top of the page, one for the accounts department and another, with a different code, for general enquiries. Customers, who were probably a bit annoyed with the letter were looking at the code, then at their phone as they typed in the digits, then back to the letter and mistakenly dialled the other number – result a call to another business who were none too happy.

The telephone numbering scheme is a visible element in a very important aspect of the telephone network – standardisation and co-operation. You expect to be able to dial a call to anywhere in the world but that means all the links in the chain (the exchanges, cables, satellites etc) must be compatible and interwork.

If you think of the difficulty you may have had in the past connecting a printer or wifi router to your PC which are supposed to be compatible then multiply this 10,000 times and you get the idea. What is more if a new version of exchange is introduced it must be compatible with all the others some of which might be 50 years old! In 1924 an organisation was set up to manage this work, it is still the bedrock on which all our networks are based and is now know as the ITU (International Telecom Union), a huge amount of work goes on between suppliers and operators of equipment to make sure it all fits together.

Chapter 11 – The White Heat of Technology

The phrase "The White Heat of Technology" was coined by Prime Minister Harold Wilson in 1963 marking the revolution taking place in the world. A bit rich from someone who, as mentioned in Chapter 7, prolonged the life of Strowger equipment to preserve jobs in his constituency. Telecoms had it's fair share of developments in the 50's, 60's and 70's, some truly revolutionary such as the first transistors, the digitization of speech, satellites connecting the world and optical fibres.

The Transistor

Up until 1947 if you wanted to control a larger electrical signal with a weaker one then the only option was a valve. These were used to amplify long distance calls and were at the heart of Tommy Flowers computer built to decode German messages.
The trouble with valves is they generate a lot of heat, use a significant amount of power and are delicate, they don't like vibration nor being turned on and off.

William Shockley of Bell labs had been working on an alternative for 10 years but, whilst understanding the principles he was unable to make it work. He was then joined by John Bardeen and Walter Brattain who managed to create the first transistor which proved Shockleys theory was sound. Shockley then went on to develop a better version which revolutionised the electronic industry.

It was not until transistors were available to work as repeaters that transatlantic telephone cables were possible, valves were too delicate and required too much power to work under the sea.

Satellites

The author Arthur C. Clarke is often quoted as being the inventor of the concept of the communications satellite which he prophesied in 1945 by suggesting a satellite positioned 22,236 miles above the equator would remain in the same position as the earth rotated and could therefore be used to bounce messages across the world. The first practical communications satellite was named Telstar and was developed by AT&T in co-operation with the British and French Post Office Telephone companies. It was launched by NASA in 1962 and was the first example of a privately sponsored space launch.

The main snag with Telstar was that it was in a low orbit of 3,600 miles above the equator so it was continually moving. It was only visible for 2 hours 37 minutes from any one point and the period it was available to carry traffic between the UK and America was different each day so TV programmes had to be arranged to make use of the satellite link at the right time. The radio dishes at the earth stations also had to move in synchrony with the satellite. The UK end was at the Goonhilly Earth station in Cornwall.

The first Telstar proved the feasibility of the technology and soon more powerful satellites were launched into geostationary orbits, 22,236 miles above the equator just as Arthur C. Clarke had predicted.
So why are we still laying undersea cables for telephone, internet and TV networks?

Answer: Delay – we might think a radio signal is instantaneous but in practice it takes ¼ second for a signal to travel to a geostationary satellite and return to earth (44,472 miles, ten times the distance of a transatlantic undersea cable). In a telephone conversation you can just notice it however if you need two consecutive satellite links say in a call from the UK to Barbados via America then the ½ second delay becomes a real barrier to conversation.

In some circumstances a satellite connection is the only option but most telecoms operators would rather use cable especially as fibre optics have reduced the cost of cable considerably. Internet users would also be annoyed at the delay especially if they were playing games.

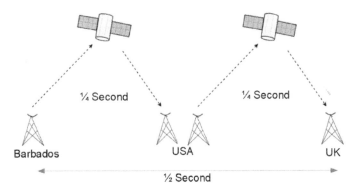

Another issue with satellites in countries which can suffer high winds (like Barbados during hurricanes or tropical storms) is that the earth station dishes have to be "stowed", that is moved into a position of least wind resistance, as they are like big upturned umbrellas. When stowed they are not pointing at the satellite so service is lost.

Whilst not part of the fixed telephone network an interesting development in the 1990's was the Iridium satellite constellation which provides voice and data connections to satellite phones, pagers over the entire Earth surface. It was developed by Motorola and came into service in 1998.

The constellation consists of 66 active satellites in orbit, required for global coverage, and additional spare satellites to serve in case of failure. Satellites are in low Earth orbit at a height of approximately 485 miles, close enough to the ground that hand held equipment can be connected. This system is used in remote areas and by news organisations.

A similar system has now been launched by SpaceX with low orbit satellites 210 miles above the earth delivering internet connections throughout the world.

Incidentally, it was not until 1923 that the first transatlantic telephone call was made, the distance was too far for a cable, it used wireless. Calls cost £15 for three minutes, over £900 in today's money! Mind you there was only one circuit so it is hardly surprising it was expensive. The first transatlantic telephone cables were laid in 1955 and 1956 (TAT1, two cables one in each direction with 16 transistor repeaters) and each cable had 35 speech channels. There are now over 30 transatlantic cables.

Digital

One of the biggest contributions to today's technology by the telephone engineers researching better ways to carry calls is something with the snappy title of Pulse Code Modulation or PCM which tends to be referred to as "digital".

Earlier in this book I mentioned the ambition of many of the early inventors was to be able to carry more than one telegraph message down the same wire. In 1853 an American called Moses G. Farmer suggested a system which sampled each dot or dash only sending a sample down the line and then sending another sampled message in between times and in 1903 another inventor, W. M. Miner, actually made a working system. But getting this to work for speech is rather more complicated.

When a sound is made the diaphragm in the microphone moves in and out with the vibrations in the air. The aim of the telephone, gramophone, CD player or radio is to capture these vibrations in an electrical signal carry them across the network or record them on a disk or player and then cause diaphragm in the speaker to move in the same way thereby replicating the sound.

If the initial signal is then carried over a long pair of wires by the time it reaches the far end it is less strong.

The obvious answer is to amplify it but every time this is done a little bit of noise or distortion is added so a call being made from London to Sydney might be amplified many times and each time gaining a bit more distortion. Those of us old enough to remember cassette tapes will also remember friends passing us tapes to copy so we could listen to a record for free. The problem was every generation of copy was a bit more distorted so by the time you got a copy of a copy of a copy the sound was awful. This is exactly the same issue which for many years meant international calls were often poor quality.

So far in this book I have avoided any technical descriptions or jargon but this idea is so fundamental to almost everything we use today I make no excuse for giving a very brief description of the principles. If you know it skip a couple of pages, if you don't read on and be amazed.

In 1937 British engineer Alec Reeves who was working in France came up with the idea of converting the signal to a series of binary numbers (just 1's and 0's) which can then be sent to the other end via as many amplifiers as needed and then the numbers are converted back into a variable signal at the distant end. Just like Morse's telegraph which just sent on off signals which could be amplified without losing the message so digital messages would also be distortion free.

Reeves gained patents in 1938 (France) and 1943 (The USA) by which time WWII was in full swing and Reeves had started working at the UK Telecommunications Research Establishment.

His idea was to sample the signal at regular intervals and instead of sending the signal send a number to represent each sample and at the destination the equipment reconstructs the wave from the binary numbers.

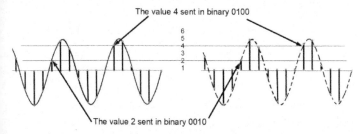

Converting an analogue signal into binary numbers is what we mean by "Digital" and is at the heart of all of today's communication – music, telephones, radio, TV, the Internet are all digital and all based on the same technology thanks to Alec Reeves.

Of course a real voice or piece of music does not make a nice regular patten as in my example above and therefore changes in the patten may get lost between the samples. The more times a sample is taken the less risk of this happening but the more data has to be sent. In telephony conversations are sampled 8,000 times a second and instead of my 6 steps above there are 256.

CD's use exactly the same technique but to maintain the high quality the music is sampled 44,100 times a second and there are 65,536 steps.

You will notice that in the drawing above there is a gap between each sample. The next clever trick was to send another conversation down the same cable in the gap, in fact not just one other conversation but another 29! So by mixing (or multiplexing to give it a fancy name) 30 speech conversations can be sent digitally down the same cable and being digital it can be switched and amplified without changing the quality of the sound. These 30 channel systems are the building blocks of modern telecommunications, for those who are interested this requires a data speed of 2Mbs.

Being sent as a series of 1's and 0's also allows the system to check for errors. For example the number of 1's sent can be counted and at the end of a block of data this count sent on. The receiving end can check it received the same number of 1's and if not ask for a repeat or flag a fault. When you play a CD and the player starts to make a staggering noise this is because the bit count has shown an error and the player tries to read the data again.

Alec Reeves was awarded a CBE for his wartime work designing a system to help bombers find their targets and PCM was commemorated on a postage stamp in 1969 but it is small thanks for someone who brought us into the digital age.

Microwave

Having to run copper cables to link cities is expensive so for many years engineers looked to use radio as a low cost alternative. The first transatlantic telephone calls were made using radio but it was prone to interference and the number of calls which could be sent simultaneously was small.

A good alternative would be to use what are known as microwaves, these can be directed in straight lines using a dish at each end of the link. In 1931 an Anglo-French consortium managed to operate a link across the English channel which successfully carried telephone traffic although the technology was not really good enough for it to become a practical solution. Radar also uses microwaves and its development during WWII had a knock on effect of providing the components needed to build a practical microwave network.

The main difficulty with microwave is it relies on having a line of sight between the two points which are connected. So in the 1950's and 60's many tall structures were built in city centres and on hilltops between them to create a microwave network. The most famous of these is the Post Office Tower which at the time was the highest building in London and it had to be, any building in its way would block the signals.

The network was also a key part of the early warning system during the cold war, anyone taking a close look at the maps published by the GPO of their microwave network would be surprised to see connections to the middle of Norfolk and the North York Moors, not renowned centres of telecommunications.

The network was also used to relay TV pictures around the country, the BT Tower was the best place to watch the football on Saturday afternoons as all the matches being filmed around the country were fed back to the London studios via the tower.

Microwave is no longer used for the core of the network as fibre is more reliable and efficient but it is still used for short distances such as connecting mobile phone aerials to the network or to link islands to the mainland. Microwave technology is also used to connect to and from satellites. The BT Tower is no longer needed for microwave but it is now a listed monument so it won't be knocked down.

My greatest achievement in my BT career was when I hosted a meeting with a media company, we were allowed to use the BT Tower revolving restaurant as a venue. Having started the proceedings I realised we were not going round, a call was made and suddenly there was a jolt and off we went. A career highlight I am sure you agree although regrettably we were only given sandwiches for lunch.

Optical Fibres

Optical fibres have been around for many years, they were used in bundles to give doctors a picture of the inside of patients throats and lungs. The first use of such a device was in the 1930's and in 1956 a system was patented by researchers in Michigan.

From a telecoms point of view the fibre optic cable only came about with the development of lasers to produce a light in which data can be carried and the losses were low enough that the light would travel a long way before needing amplifying or a repeater.

The biggest breakthrough came in 1965 when a couple of scientists working for Standard Telephones and Cables in London managed to reduce the loss in fibres by using silica such that they could carry messages for 90 miles between repeaters. Charles Kao won a Nobel prize for this work.

The American company Corning Glass carried on the development and now fibre cables are much cheaper and can carry more data than copper.

They are also thinner and therefore take up less space in the ducts under the roads and pavements, the only downside is that they need electronics at both ends. The copper wires to your home just need a socket, a fibre needs a box of electronics and power, it won't work in a power cut unless there is a battery backup.

hn Bardeen, William Shockley and Walter Brattain of Bell labs who developed the first transistor

Alec Reeve, the inventor of PCM, everything we listen to or watch on TV today relies on his invention

Chapter 12 – Competition

The Bosto Times 1876: *"A fellow can now court his girl in China as well as in East Boston; but the most serious aspect of this invention is the awful and irresponsible power it will give to the average mother-in-law, who will be able to send her voice around the habitable globe."*

In 1887 the Post Master General told Parliament *"having regard to the cheap and swift means of communication which at present exist by means of the telegraph between the principal towns in the UK it is extremely doubtful whether there would be much public advantage in establishing telephonic communication generally between those towns"*

Selling the first phones was difficult as you needed two phones, your own and someone to call. In the 1990's Deutsch Telecom were trying to launch an ISDN video phone and had to sell them in pairs. This was 25 years ago and it was seen as a potential new market, now we just use skype for free.

A telephone was a very expensive luxury, £20 a year in 1900, not much different from the cost of a servant who was probably rather more useful. By the start of WWI there were only 1.7 phones per 100 people in the UK opposed to 9.7 in the USA and even 4.6 in New Zealand.

Up until the 1930's the ethos of the Post Office was one of a Government Department which graciously allowed people (Subscribers, not Customers) to use the telephone. A newspaper advert appeared:

The Friendliness of a Phone

It's a wonderful pal. Are you lonely? A ring will put you in touch with friends. Nothing doing tonight! Prr-ing! And invitation to the theatre or 'Come on over and make a four at bridge' Daddy lost the last train! Well, anyway, it's good to hear his voice – to know he is all right – to have a car at the junction.

This advert was not paid for or promoted as you might think by the Post Office who would benefit directly but by the organisation representing the equipment manufacturers who wanted the Post Office to buy more equipment!

A new Post Master General appointed in 1931, Kingsly Wood, changed that by firstly negotiating a financial separation from the treasury (up until then any surplus cash when straight to the government so there was little incentive to expand the business) which allowed more investment in automatic exchanges and then he set about advertising the service which in turn generated more revenue. He also introduced cheap rate calls in the evenings and weekends. He championed publicity, he was the force behind the famous Post Office mail films of the 1930s.

But in 1939 the war changed things and put a huge strain on the network, social use was discouraged and even regarded as a bit suspicious.

After the war the telephone started to move from a luxury item to something middle class families wanted. It was useful, but also a status symbol, you were so important that people needed to contact you, the phone was put in the hall so visitors could see you were connected. Another reason for it being in the hall was it discouraged long conversations, people kept calls to a minimum to avoid unexpected bills and making a trunk or international call was not something you did without good reason.

Talking of unexpected bills, the subscriber, apart from keeping their own record, had no way of knowing in advance how much their bill would be. Charging was done in "units" so a local call might be three minutes for one unit whilst a trunk call might only be 30 seconds. Each time a unit was spent an electro mechanical meter in the exchange moved on one step.

The subscriber was sent a bill every three months with the reading of their last bill, the new reading and the difference which was then used to work out the cost. These meters, hidden away in the exchange where the customer could not see them, were photographed every three months and the photograph became the permanent record of the number of units used.

There were all sorts of disputes when people suddenly got unexpected bills, perhaps because a teenage child was ringing America or because of a fault. It further reinforced the impression of the Post Office or BT being a law unto themselves.

Photographing the Meters.

Each subscribers line had an electromechanical meter in the exchange. The accounts department used to photograph to determin the units used since the last time they were photographed (every 3 months)

Another way to encourage phone use was the introduction of "Yellow Pages", telephone directories arranged in business type order, so all the Florists appear together in one section, the Plumbers in another. So if the customer wants a plumber in a hurry they can see a list of all the candidates quickly and easily. It was a great money spinner for the telecoms companies as to be listed the business had to pay a fee, the more you paid the bigger the print used and pay even more and you could also have an advert alongside. If your business was not listed you would lose out to your competitors. The UK Yellow pages was started in 1966 (Hull had one earlier), the printed version was finally killed off by the internet in 2019.

It was suggested that the telephone also got a boost from the Street Offences Act of 1959 which took prostitutes off the streets and onto the phone!

In the 1960's and 70's there were many areas where there were not enough cables in the ground so that many prospective customers had to go onto a waiting list. To increase the utilisation of the cable network shared service phones were introduced (called party lines in the USA) where two customers used the same pair of wires. To make a call you pressed a button on your phone which momentarily put an earth condition on one wire to indicate which caller was making the call. You then made the call as normal but if the other customer picked up their handset they could hear everything you said.

On normal telephones the ringing current (to sound the bell) was sent over the two wires, for shared service it was sent over one wire and the other side of the bell was connected to earth. So one phone was rung on wire A and the other on wire B. If the other customer was using the phone you could not receive a call. My first phone in 1975 was shared with the house next door who had three teenage children so the line was always busy. Shared service customers got a reduction in their bills for this inconvenience. Shared service was introduced in 1942 and was not abolished until the early 80's.

It's good to talk – there were waiting lists in some areas for a line in the 60's and 70's so all the marketing was aimed at getting those who had a phone to use it more rather than disappointing potential new customers. Busby, a cartoon bird, was used in one major advertising campaign and the cost of calls was reduced further in evenings and at weekends as, in theory, the network is not congested outside business hours.

This is not the case at Christmas when everyone wants to call their relatives, particularly in places like Australia, the Caribbean and Canada. The snag is that as these are not major business destinations there are fewer international circuits to them and the result is considerable congestion on Christmas Day.

It is at times like this that every international operator changes their routing configuration. For example the Japanese don't celebrate Christmas, to them it is a normal day but all the circuits to and from Japan are empty so they are able to sell their capacity to Australia to the highest bidder. It always seemed daft to me that BT made Christmas Day international calls cheap rate, they ought to have made them full price but offered a big discount for Christmas Eve and Boxing Day to spread the load and make it easier for callers who would get through more easily.

In the 1960's and 70's the Post Office telephone service had a terrible reputation for quality of service, like most nationalised industries it was an easy target for the press and the old fashioned Strowger exchanges did nothing to improve the image. Other countries had push button telephones, but the UK still had rotary dials and the Post Office owned the telephone instruments, the subscriber (not customer) just rented them, if it became faulty an engineer came round to fix it. The phones were very heavy and not stylish and if you wanted an extension that was extra rental on your bill each quarter. It was not until privatisation that the phone sockets were added and you could fit your own phone. Up until then it was felt that the subscriber was not to be trusted with connecting their own equipment to the network although it was difficult to see what damage they could do, the equipment at the exchange end was a fairly robust collection of relays. If anyone was prosecuted for putting illegal equipment on the end or tapping someone's phone they were charged with the rather strange offence of "stealing electricity".

Because the network was unreliable BT encouraged companies to build their own private networks which were connected together by BT circuits as they were the only people allowed to do so. These private circuits were very costly and therefore a good revenue stream for BT and at the same time most large companies had their own large telecoms departments who managed their internal networks. By 1990 the UK had many thousands of private PCM circuits, far more than any other country in the world, Spain for example had less than 20.

Things started to change when in 1969 the telephone part of the Post Office was separated from the rest of the organisation (letters and Post Offices) and became a Public Corporation. Then in 1981 it was separated completely becoming British Telecom (and then BT).

The UK's overseas telecoms interests were managed by Cable & Wireless, a Government owned company who had their own cable ships and ran the telecoms networks in Hong Kong and many Commonwealth countries particularly in the Caribbean. As part of the Thatcher governments privatisation programme both Cable & Wireless and BT were floated on the stock market and given much more freedom to manage their own affairs.

This freedom was accompanied by the opening up of the telecommunications network to rivals the first of which to get their own UK licence was Mercury Communications who were owned by Cable & Wireless and funded initially by BP and Barclays Bank although their shares were soon bought out by Cable & Wireless.

Mercury were only interested in the long distance and international markets. They built a fibre network which was run along the sides of the railways which connected together a number of System X exchanges. Important large customers had direct connections to the Mercury exchanges but anyone else was given a "smart box" which in practice dialled 131 followed by a 10 digit account number which allowed them to route calls from their BT phones into the Mercury network. Of course with 98% of all telephones still connected to BT almost all calls were routed through the Mercury network and then connected to the local BT exchange at the far end. The biggest beneficiaries were the large companies with their own telecoms networks who could easily divert all their long distance traffic between the rivals but the threat of competition also put a limit on the cost of trunk calls.

By then from an engineering point of view the cost of a long distance call was not much different from a more local one, BT could have slashed it's long distance prices at a stroke and killed this usurper stone dead but the government would have not allowed this blatant monopoly behaviour.

Soon afterwards other companies joined in the fun, cable TV companies started digging up the pavements in residential streets offering telephone services on the same cables as TV and eventually broadband but in the 1980's that did not exist. Perhaps a bigger threat to BT was that in city centres companies like Metro Fibre Systems (MFS) and City of London Telecom (COLT) ran fibre cables round the city centres to cream off the big telecoms users. They then either routed the calls to other big cities themselves or passed them to Mercury. At one point BT had lost 50% of the central London business traffic.

It took a long time for BT and their competitors to establish the engineering rules and processes to allow the interworking at every exchange, many in BT tried to be as difficult as possible to the new competitor, at the time I had a role in managing BT's traffic and I had many arguments with colleagues who could not see that we had to co-operate. If someone with a BT phone cannot call a number which happens to belong to Mercury they won't realise where the problem is and assume it is BT's fault, that's who they pay their money to. They may actually switch to a competitor because of problems caused by Mercury.

The speed with which Mercury were able to build their network was largely due to the deal with Network Rail who allowed them to run their fibre cables alongside the railway tracks in concrete troughs. No need to dig trenches and put in expensive ducts but this also left the cables exposed to vandalism. In the past a favourite trick of criminals was to pull out a great length of cable, run it over the tracks so that the next train would cut the cable which could then be carried off for scrap. This was no good with fibre cable which has no scrap value but, particularly during the school holidays, teenagers would try the same trick and on several occasions in the early

1990's Mercury lost a huge amount of capacity very suddenly.

Because of the modernisation programme, the ability for customers to buy their own instruments and have many phones in the house together the reduction in cost meant that the telephone was no longer on a shelf in the hall but was in the kitchen and in the living room. One side effect of this was the sudden rise in TV phone-in programmes where callers were encouraged to call a number in order to win a prize, each call was charged at a premium rate with a portion of the revenue going to the TV Company. The call levels were surprisingly high and a great deal of money was made simply because people no longer saw a telephone call as expensive and it allowed instant participation.

I was involved with supplying equipment to Telia, the Swedish telecom company who supported a TV Programme called Bingo Lotto. Each week people bought bingo cards from shops and newsagents and on Saturday night watched the programme carefully crossing the numbers off each game. If they got a complete line of numbers instead of shouting Bingo! they called a special number. The odds are arranged that thousands of people get a line at the same time, they all ring the number and all but one get an announcement saying well done, better luck next time and are charged about 25p. If they take the card to the newsagent the next week they win a small cash prize of about £10 but the lucky person who got through to the studio won a car, a holiday or a large amount of money. Several games are played during the course of the programme. No one expected the game to be quite as successful as it was, I was told by several different people when I mentioned I was involved that the host, a very personable man called Leif Olsson, was offered a very small percentage of each card sold as a fee or a salary. He did not think it would be a big success and took the salary, apparently if he had taken the percentage on each card he would have become the richest man in Sweden.

In the mid 1990's Bingo Lotto attracted 4.5 million viewers to every programme, the highest viewing figure that has been registered in Sweden, a country of only 10.5 million people.

Now we use the internet to interact with others quickly and the days of the phone-in programme have passed with perhaps the exception of the Children in Need or ITV Telethons.

There is also a very negative impact of phone-in's, that is when there is a disaster people feel duty bound to try and call their loved ones in the area where the problem has occurred. When this happens the telecoms companies have to work together to quickly block the calls at source so as not to overload the local network just when the hospitals and emergency services need the telephone the most. The Hillsborough disaster in 1989 was a UK example. At the start of football match in Sheffield 96 Liverpool fans were crushed to death by the pressure of too many people in the ground. The news of the disaster was broadcast immediately as it was a big match and thousands of families in Liverpool started calling all the numbers they could find of hospitals etc in Sheffield. There were millions of calls, which if allowed to get to Sheffield the local exchanges would have become overloaded and failed after which no one would be able to call emergency services and the hospitals be unable to call for reinforcements, order blood or call for any other support. Fortunately the BT managers in their control room managed to block the majority of calls to Sheffield from leaving the Liverpool area and the Sheffield exchanges kept working.

A similar although much bigger event happened in San Francisco during a major baseball event in the same year, the game was about to start, the TV cameras were on when the picture started to shake, the last word the commentator said before the screen went blank was "Earthquake". You don't have to have a Phd in geography to know that San Francisco and earthquakes don't mix and almost everyone in the rest of the USA who knew someone in San Francisco picked up the phone and called them. Hundreds of millions of call attempts were made to an area which was suddenly in need of every one of its telephones. Again, fortunately the incoming calls were quickly killed protecting the local network.

Chapter 13 – The Mobile and Internet Revolution

This book is about the fixed telephone network but it is perhaps important to describe a little of the history of this upstart mobile technology which may well make it obsolete.

Up until 1985 a mobile phone was in practice a two way radio link with a base station, the sort of thing the police, fire brigade or taxi's used. There was a Post Office car radio telephone service in London but this required a big transmitter/receiver and for most of the time it was in operation it required an operator to connect the calls. It was so expensive only the very rich could afford it.

The Scandinavians were the first to see the possibilities of a cellular mobile phone network. They formed the Nordic Mobile Telephone Group in 1969 to work on the technology and their specifications were adopted as the European standard in 1982.

The British Government awarded two licences in 1983 to use a fairly restricted radio spectrum for this new service from 1st January 1985. One licence was awarded to Racal Telecom, an electronics company who specialised in radar and wireless communications systems for the armed services, ships and aircraft. There was a rumour that when the Racal board realised how much investment they had committed to build the first mobile network they were appalled, their annual profits were around £100m and these were put at considerable risk by this mad cap plan so they sacked the director responsible.

The other licence was won by BT but as the government were concerned that BT would use it's monopoly position to stifle competition they needed a partner. Securicor, the company who deliver cash to shops and banks, joined them with the rather vague connection that they also had a wireless network to manage their cash deliveries, it was a marriage of convenience in which they formed a company called Cellnet, 60% owned by BT and 40% by Securicor. Securicor played no part in building or managing the network and contributed the princely sum of £4m, the best investment they ever made.

The first public mobile call was made by the comedian Ernie Wise from outside the Dickens' Pub in St Catherine's dock to the Racal HQ in 1985. For some reason he was dressed in Dickensian clothes, I am not sure what that had to do with telephones of any sort.

Initially, most mobile phones were still in-car dashboard-fixed handsets. The technology was still too big to be carried by hand. In 1987 it cost £1,695 to install a BT Cellnet car phone (about £4,800 at today's prices) and the call charges were extortionate. I was working for BT and had shared use of a Motorola hand held phone, shared because it was so expensive that we had to share it. The handset cost several thousand pounds, was the size and weight of a bottle of wine and the battery lasted about 30 minutes if you were lucky, that is when I said "**It will never catch on**".

It was thought that no one would want such costly toys (sound familiar?) and a couple of companies offered Telepoint services as an alternative. They provided small handsets which linked to access points in stations and other public places, rather like WiFi today, the largest such network was branded as Rabbit, presumably from the Cockney rhyming slang, rabbit and pork – talk. You could only make outgoing calls and had to stay within 100m of the base station so in practice it was only good for calling home or a taxi from the station on your way home. Eventually there were 12,000 Rabbit base stations and 10,000 customers. It closed after 20 months! The service was run by Hutchison, a Hong Kong based telecoms company. After the spectacular failure of Rabbit they went on to bigger and better things, they won a mobile licence the next time they were offered and instead of Rabbit called their company Orange.

As we all now know mobile phones became smaller, cheaper and more sophisticated and took over the world! Racals network was branded Vodaphone and very quickly grew into one of the largest UK companies, so large that Racal split itself from it's precocious child and went back to making systems for the defence industry. A similar problem occurred for Securicor, their stake in Cellnet was many times bigger than the value of Securicor itself and in 1999 BT bought out their share for £3.15 billion, so it was not a bad £4m investment over 16 years for no effort.

There were several upgrades and changes to the standards used to improve the quality of service, to increase the number of available handsets, the compatibility between countries and to add data connections. In 1992, just 20 years ago, a developer working for Vodaphone was asked to create a messaging system, his name was Neil Papworth aged 22 at the time and he sent the very first text message "Merry Christmas" to a director at Vodafone, who was at the office Christmas party.

Another major step forward in mobile phone use came about in 1996 with the introduction of Pay as You Go. This took off when, after the first two operators had been given a chance to establish the market, two new players were licenced; Mercury One to One, an offshoot of Mercury the fixed network carrier and Orange, set up by

Hutchison in 1994. Pay as You go opened the market to youngsters who were less likely to want or be given a standard contract and they fell in love with messaging as it was much cheaper than making calls.

Just how valuable the mobile phone business had become was evident in April 2000 when the UK government auctioned the radio spectrum's needed to run the next generation of mobile phone known as 3G. There were five licences available, Vodaphone bid the highest, £5,947 million pounds, that is £100 for every person in the UK most of whom in 2000, did not possess such a new fangled thing as a mobile phone. The other four successful companies paid over £4,000 million each.

So by the new millennium people were already announcing the end of the fixed line. Why do you need it when a mobile phone allows you to do exactly the same but anywhere you want, why use a house phone? It was still cheaper to use a fixed line but as mobile calls and rentals became more affordable that advantage began to drop but then a miracle happened – The Internet.

Being British we all like to say the internet was invented by Tim Berners-Lee, a British scientist working in a lab in Switzerland in 1990. It is true he devised the idea of making hypertext documents available to anyone on a data network but all the protocols and packet switching between computers had evolved over many years by many different engineers. Just as Bell got all the glory for the

Telephone so Tim Berners-Lee got it for the World Wide Web the big difference is that he did not try to patent it and made it available to us all for free.

It took a while for the public to see the advantages of the web, Amazon was started in 1994, initially to sell books but it soon added other products. Within 10 years it was the largest retail business in the USA! But users could only access Amazon or any other web site by connecting their computers to the web servers using their phone lines. All PC's were soon equipped with dial up modems and if you were lucky you could connect at the fantastic speed of 64kb/s. But soon technology in the form of Asymmetric Digital Subscriber Line equipment, or ADSL for short, came to the rescue.

When I was at college in the 1970's if anyone suggested that transferring data at anything higher than 64kb/s down a telephone line was possible would have been regarded as mad but in 1988 a patent was awarded to Bellcore (previously Bell Labs) for technology which allowed 100 times higher speeds. At first this was an interesting lab exercise but fairly soon the equipment manufacturers were able to produce low cost equipment which made ADSL a practical proposition to increase the connecting speed between the telephone exchange and the customer. This is often called "the last mile" but in practice can be anything up to the last three miles and despite the engineers best efforts the further away you are from the exchange the slower it gets, that's just the laws of physics although when you hear some complainers you would think it is just spitefulness on the part of the telephone companies against people who don't live in towns.

So in the 1990's this technology suddenly opened up new possibilities for the telecoms companies, did they embrace the new world of the internet? No, it was all still a "toy", what interested them was the ability to take on the cable TV companies and deliver TV as well as phones. BT, Telewest, Virgin Media were all after the Pay TV market and all the discussions at the time were about rolling out this new technology and came down to an equation –

If you upgrade a street how many will take TV and at what price?

Initially there was little thought given to the idea of delivering broadband but as the 21st century approached all the talk was of "Triple Play", selling broadband, TV and telephone as a package and then Quad Play when mobile was added as well. If the customer paid £30 a month for TV, £10 for broadband then the telephone element was largely irrelevant and the telecoms company directors started to view themselves as media moguls.

The internet revolution has forced people to retain their home telephones, for the moment. There is now an arms race between the fixed line and mobile companies. The latter are pushing the next generation of mobile network called 5G which will provide much higher speeds than

ADSL but in reply the fixed line networks are moving to "fibre to the home" where you will have a fibre optic cable connecting your home to the exchange with the potential for incredible speeds. Who will win? If there is anything I have learnt from writing this book it is that predicting the outcome of any development is foolhardy. Pony Express, the Telegraph, Rabbit were all supposedly "winners" which are no longer with us. In Africa the fixed telecom network did not get outside the main towns and now mobile has effectively become the only system the public use. A few years ago we were in a remote part of Namibia trying to visit some caves, we asked a lady in the village about getting access and she went into her hut, took out her mobile phone and called the curator. The village had no electricity, running water or sewage, presumably she charged the phone when she went into town. In fact in less developed countries like Somalia mobile phones are also used as the only reliable way to transfer money so in many respects they have leap frogged Europe.

So will broadband offer the fixed network something of a lifeline or will it be it's "swan song", whatever happens in the future I expect I for one will say "**It will never catch on**".

About the Author – John Lucas

I was born and brought up in North London and joined the GPO as an apprentice in 1969 at the age of 16.

I spent 23 years with the company (which became BT) in a wide range of posts including periods in exchange construction, training, research and network management.

I then left BT and worked for a while as a consultant on an EU project in Greece, in Trinidad and Tobago advising on Network Management for new operators such as Mercury and MFS before joining Ericsson marketing department.

This was followed by a number of roles in smaller telecoms companies and consultancies. I retired in 2018.

I was planning to do a talk on the history of the telephone, it would be easy, just get a book on the subject to check on dates and away you go. I then discovered that the only book I could find was written in 1910! So I decided to write my own, I do hope you like it.

I am always pleased to hear from readers so by all means email me at john@itwillnevercatchon.com. I will try to reply within a day and look forward to hearing from you.

Printed in Great Britain
by Amazon